Wastewater Operator Certification Exam Prep

By Kenneth C. Morgan, P.E., and David L. Russell, P.E.

American Water Works Association

Wastewater Operator Certification Exam Prep

Disclaimer

Although this study guide has been extensively reviewed for accuracy, there may be an occasion to dispute an answer, either factually or in the interpretation of the question. AWWA has made every effort to correct or eliminate any questions that may be confusing or ambiguous. If you do find a question that you feel is confusing or incorrect, please contact the AWWA Publishing Group at books@awwa.org.

Additionally, it is important to understand the purpose of this study guide. It does not guarantee certification. It is intended to provide an operator, or aspiring operator, with an understanding of the types of questions he or she will be presented with on a certification exam and the areas of knowledge that will be covered. AWWA highly recommends that readers study and understand the reference material from which the questions are drawn.

Director of Publishing: William L. Murphy
Sr. Editorial Manager: Kimberly J. Retzlaff
Product Manager: Tony Petrites
Production Manager: Gillian Wink
Project Manager: Peggy Tyler
Cover Design: Michael Labruyere

Library of Congress Cataloging-in-Publication Data

Has been applied for.

ISBN: 978-1-64717-009-7

Printed in the United States of America
American Water Works Association
6666 West Quincy Avenue
Denver, CO 80235-3098
awwa.org

Contents

About AWWA, v
Preface, vii
Acknowledgments, ix

Questions

I. Wastewater Collections
Math, 1
General, 9

II. Wastewater Treatment
Math, 41
General, 47

Answers

I. Wastewater Collections
Math, 89
General, 103

II. Wastewater Treatment
Math, 147
General, 155

About AWWA

American Water Works Association

The American Water Works Association is an international, nonprofit, scientific and educational society dedicated to providing total water solutions assuring the effective management of water. Founded in 1881, AWWA is the largest organization of water supply professionals in the world.

Our membership includes over 4,000 utilities that supply roughly 80 percent of the nation's drinking water and treat almost half of the nation's wastewater. Our over 50,000 total memberships represent the full spectrum of the water community: public water and wastewater systems, environmental advocates, scientists, academicians, and others who hold a genuine interest in water, our most important resource.

AWWA also takes great pride in helping establish two preeminent organizations dedicated to safe water, the Water Research Foundation in 1966 and Water For People in 1991.

AWWA unites the diverse water community to advance public health, safety, the economy, and the environment.

What we do:

- Offer education to water professionals
- Advocate for safe and sustainable water
- Collect and share knowledge
- Create volunteering opportunities

Preface

This book was developed by the American Water Works Association (AWWA). All of the questions are based on ABC's Need-to-Know Criteria for certification of wastewater collections operators and wastewater treatment operators.

All of the questions in this book are new to this edition and have been thoroughly reviewed and vetted by subject matter experts in the water industry. Each question is labeled with the ABC certification grade: Grades 1 & 2 (introductory) or Grades 3 & 4 (advanced). Each answer provides the ABC Need-to-Know criteria specifying the area of operator knowledge covered by the question. For questions involving math, the worked-out math is shown after the answer.

This guide is intended to give operators practice in answering questions that are similar in format and content to the questions that appear on certification examinations. While the questions found on an exam are not duplicated in this guide, similar types of questions are included; they cover relevant areas of study and allow operators to understand what areas they need to study for passing an exam. Operators who have difficulty answering any of the questions in this book should consult the reference source provided (if available) following the answer to the question.

Study Tips

1. Use this book alongside other training materials and/or courses, not in place of them. The questions in this book are designed to test your knowledge and identify content areas in which you need to complete further study. Testing yourself with the questions in this book is an excellent way to review your understanding of a topic.
2. If you are having trouble with a question or set of questions, consider consulting the specific reference mentioned after the answer for further study.
3. Don't memorize the questions and answers. Questions in this book will not be duplicated exactly in a certification exam. If you understand the context of a question, you should be able to answer any number of similar questions on the topic.
4. Check with your local certification agency (state or provincial) so you know what subjects and levels will be covered under your specific certification examination. There are significant regional and state/provincial variations in regulatory requirements, certification criteria, treatment plants, and distribution systems.
5. Give yourself as much time as you can. Studying works best when you have the time to review material several times.

References and Resources

The following publications were consulted during the writing of this book, and are referenced after the question answers in this book.

Operation of Wastewater Treatment Plants, Volume 1 (7th ed., 2008, Sac State OWP)
Operation of Wastewater Treatment Plants, Volume 2 (7th ed., 2007, Sac State OWP)
Operation and Maintenance of Wastewater Collection Systems, Volume 1 (8th ed., 2018, Sac State OWP)
Operation and Maintenance of Wastewater Collection Systems, Volume 2 (7th ed., 2010, Sac State OWP)
WSO Basic Science Concepts and Applications (4th ed., 2010, AWWA)

Acknowledgments

The following individuals served as volunteer reviewers for the questions in this book. Their service and dedication to the water operator profession and the water sector is greatly appreciated.

Ari Copeland, PO (Black & Veatch)

Robert Cheng, PE, PhD (Coachella Valley Water District)

Clifford Fletcher (Anchorage Water)

Emma Larson (City of St. Cloud, MN)

Shaun Livermore, PO (Poarch Band of Creek Indians Utilities Authority)

Pete Peterson (Milwaukee Water Works)

Daniel Pfleiderer, MBA (Florida Gateway College)

Damon Roth, PE, BCEE (Brown and Caldwell)

Robin Rupe (Northeast Ohio Regional Sewer District)

The authors would also like to thank the following individuals who helped improve the clarity and accessibility of the questions and answers in this book: Todd Brewer, Tony Petrites, Kimberly J. Retzlaff, Somlynn Rorie, and Melissa Valentine.

This page intentionally blank.

I. Wastewater Collections, Questions

Math

1. What is the average velocity (in feet per second) of a particle that travels 1,200 feet in 5 minutes?

   a. 100.0 ft/s
   b. 4.0 ft/s
   c. 240.0 ft/s
   d. 3,600.0 ft/s

2. How far can a particle travel (in feet) if it is suspended in a volume moving at a rate of 1.5 ft/s for 9 min?

   a. 60 ft
   b. 360 ft
   c. 90 ft
   d. 810 ft

3. Convert 2.25 mgd to cubic feet per second.

   a. 3.49 ft^3/s
   b. 26.04 ft^3/s
   c. 1.45 ft^3/s
   d. 1.55 ft^3/s

4. How long will it take (in minutes) for an object to travel if it is moving at a rate of 2 ft/s over a distance of 1,000 ft?

   a. 16.67 min
   b. 33.33 min
   c. 8.33 min
   d. 21.00 min

5. Determine the flow rate (in gpm) if the flow rate is 1.2 ft^3/s.

   a. 448.80 gpm
   b. 493.68 gpm
   c. 538.56 gpm
   d. 583.44 gpm

6. What is the flow rate (in gpm) of an 8-in. diameter pipe with a velocity of 2.2 ft/s?

   a. 71.40 gpm
   b. 110.53 gpm
   c. 345.60 gpm
   d. 443.23 gpm

7. Determine the velocity of a 6-in. diameter pipe with a flow rate of 0.5 ft^3/s.

   a. 0.02 ft/s
   b. 0.10 ft/s
   c. 0.39 ft/s
   d. 2.55 ft/s

8. What diameter pipe in inches will yield a velocity of 2.5 ft/s and a flow rate 1.35 ft^3/s?

   a. 8 in.
   b. 9 in.
   c. 10 in.
   d. 20 in.

9. Determine the flow rate of a rectangular channel 6 ft wide at a depth of 2 ft at 1.75 ft/s.

   a. 3.5 ft^3/s
   b. 10.5 ft^3/s
   c. 12.0 ft^3/s
   d. 21.0 ft^3/s

10. Identify the depth of a rectangular channel 5 ft wide experiencing a flow rate of 39.38 ft^3/s with a velocity of 2.25 ft/s.

    a. 3.50 ft
    b. 7.88 ft
    c. 17.72 ft
    d. 87.50 ft

11. What is the pressure (in psi) at the bottom of a tank with a depth of 25 ft of water?

    a. 10.83 psi
    b. 22.00 psi
    c. 25.20 psi
    d. 57.74 psi

12. What is the force exerted on the bottom of a rectangular tank that is 25 ft wide, 50 ft long, with a depth of water of 10 ft?

    a. 623.52 lb
    b. 5,412.50 lb
    c. 12,500.00 lb
    d. 779,400.00 lb

13. What is the pressure at the bottom of a tank 30 ft in diameter at a height of 80 ft of water?

    a. 7.98 psi
    b. 13.00 psi
    c. 34.64 psi
    d. 184.80 psi

14. Convert 70 degrees F to degrees C

    a. 21.10 degrees C
    b. 38.90 degrees C
    c. 56.67 degrees C
    d. 68.40 degrees C

15. Convert 5 degrees C to degrees F

    a. 9.00 degrees F
    b. 35.00 degrees F
    c. 41.00 degrees F
    d. 66.60 degrees F

16. What is the pump horsepower if a flow rate of 750 gpm is pumped to a height of 75 ft with an efficiency of 80 percent?

    a. 0.24 hp
    b. 11.36 hp
    c. 14.20 hp
    d. 17.76 hp

17. What is the flow rate of a pump that reduces the height of water in a rectangular tank 5 ft with the following dimensions over a period of 4 min, 15 s?

    Length: 20 ft
    Width: 10 ft

    a. 0.78 ft^3/s
    b. 1.03 ft^3/s
    c. 3.92 ft^3/s
    d. 4.17 ft^3/s

18. The upstream invert elevation of a manhole is 250.47. If the downstream manhole invert elevation is 248.95 approximately 750 ft away, what is the slope of the section of pipe?

    a. 0.002
    b. 0.004
    c. 0.006
    d. 0.332

19. What must the upstream invert elevation be at a distance of 500 ft to maintain a 2 percent slope if the downstream manhole invert elevation is 315.39?

    a. 305.39
    b. 310.41
    c. 315.41
    d. 325.39

20. Determine the force (in lb) against an 8-in. diameter plate with a gauge that reads 5 psig.

    a. 1.75 lb
    b. 252.00 lb
    c. 378.68 lb
    d. 36,173.00 lb

21. A circular clarifier has a circumference of approximately 549.5 ft. What is the surface area of the clarifier?

    a. 96,163 ft^2
    b. 48,081 ft^2
    c. 30,625 ft^2
    d. 24,041 ft^2

22. What is the diameter of a circular tank with 1,500,000 lb of water with a depth of 8 ft?

    a. 55 ft
    b. 62 ft
    c. 69 ft
    d. 175 ft

23. A round clarifier with a radius of 30 ft and a depth of 8 ft has a flow rate of 1.5 mgd. What is the detention time?

    a. 0.34 h
    b. 1.50 h
    c. 2.64 h
    d. 3.02 h

24. What is the flow rate of a clarifier (in mgd) that has a diameter of 55 ft and a depth of 7.5 ft if the detention time is 1 h, 45 min?

    a. 0.24 mgd
    b. 0.91 mgd
    c. 1.83 mgd
    d. 3.12 mgd

25. How long will it take to empty a clarifier for repairs with the following information:

    Tank diameter: 75 ft
    Tank depth: 7 ft
    Pump rate: 3.1 ft^3/s

    a. 0.40 h
    b. 2.77 h
    c. 6.92 h
    d. 26.62 h

26. If 30 lb of chemical are added to 50 gal of water, what is its percent strength (by weight)?

    a. 6.71 percent
    b. 7.20 percent
    c. 7.75 percent
    d. 10.70 percent

27. Calculate the average daily flows at a lift station with the following data:

    Monday: 1.25 mgd
    Tuesday: 1.29 mgd
    Wednesday: 1.18 mgd
    Thursday: 1.20 mgd
    Friday: 1.22 mgd
    Saturday: 1.28 mgd
    Sunday: 1.10 mgd

    a. 0.82 mgd
    b. 1.07 mgd
    c. 1.22 mgd
    d. 1.42 mgd

28. A contractor has to replace approximately 137 ft of 8-in. sewer main. How many sections of 18-ft pipe will be needed?

    a. 7 sections
    b. 8 sections
    c. 10 sections
    d. 17 sections

29. How many feet of drop are in 700 ft of 6-in. sewer pipe with a 0.035 ft/ft?

    a. 12.25 ft
    b. 20.00 ft
    c. 24.50 ft
    d. 245.00 ft

30. Determine the chlorine dosage of a wastewater treatment plant if the demand is 7.9 mg/L and the residual is 1.1 mg/L?

    a. 6.80 mg/L
    b. 7.35 mg/L
    c. 8.45 mg/L
    d. 9.00 mg/L

31. What is the volume of soil (in cubic yards) removed from a trench with a width of 4 ft, a depth of 6 ft, and a length of 1,500 ft?

    a. 888.89 yd^3
    b. 1,333.33 yd^3
    c. 1,678.67 yd^3
    d. 2,000.00 yd^3

32. What is the volume (in gallons) in a tank with a diameter of 50 ft and a total depth of 10 ft with 8 ft of water in it?

    a. 37,400 gal
    b. 100,000 gal
    c. 117,436 gal
    d. 146,795 gal

33. What is the cycle time of a wet well with a storage volume of 2,000 gal if its pump capacity is 75 gpm and the wet well inflow rate is 40 gpm?

    a. 27 min
    b. 57 min
    c. 17 min
    d. 50 min

34. What is the volume of a wet well if it has a cycle time of 45 min, with a pump capacity of 65 gpm, and an inflow rate of 32 gpm?

    a. 1,440 gpm
    b. 2,925 gpm
    c. 1,485 gpm
    d. 4,410 gpm

35. Calculate the performance indicator for a collection system with the following information:

    Length of sewer system: 2,000 miles
    Annual expenditures: $2,750,000.00
    Annual number of stoppages: 145

    a. 96
    b. 130
    c. 110
    d. 9.5

36. What is the performance rating of the wastewater utility with the following:

    Performance indicator: 125
    Assumed standard: 90

    a. 1.38 percent
    b. 0.72 percent
    c. 139 percent
    d. 72 percent

37. A worker installed a plug in an 8-in. diameter sewer main and was able to test it up to a pressure of 10 psi. What was the force (in lb) against the plug?

    a. 778.0 lb
    b. 5.4 lb
    c. 502.4 lb
    d. 5.0 lb

38. What does the force (in lb) need to be against a 12-in. plug to achieve a test pressure of 8 psi?

    a. 14 lb
    b. 904 lb
    c. 6 lb
    d. 90 lb

39. A 5 ft diameter upstream manhole is plugged downstream. The elevation in the manhole drops 8-in. within 1-1/2 h. What is the amount of water loss due to exfiltration in gallons per day?

    a. 782 gpd
    b. 65 gpd
    c. 391 gpd
    d. 1,566 gpd

40. At a wastewater treatment plant, the influent biological oxygen demand (BOD) is 240 mg/L with an effluent BOD of 30 mg/L. What is the plant's efficiency for BOD removal?

    a. 87.50 percent
    b. 1.13 percent
    c. 7.00 percent
    d. 0.88 percent

41. What needs to be the effluent biological oxygen demand (BOD, in mg/L) to achieve a removal efficiency of 85 percent if its influent BOD is 270 mg/L?

a. 47.6 mg/L
b. 230.0 mg/L
c. 40.5 mg/L
d. 68.5 mg/L

42. An agency has a chlorinator set at 15 lb of chlorine per day. With a flow of 700 gpm, what is the dosage in mg/L?

a. 14.85 mg/L
b. 1.78 mg/L
c. 0.18 mg/L
d. 5.60 mg/L

43. What is the chlorine feed rate (in lb/day) of a system with a flow of 500 gpm with a dosage of 2.8 mg/L?

a. 16.81 lb/day
b. 48.38 lb/day
c. 0.28 lb/day
d. 23.35 lb/day

44. What is the flow rate in a rectangular channel with a velocity of 2.2 ft/s that is 2.5 ft wide at a depth of 1.5 ft?

a. 5.50 ft^3/s
b. 1.32 ft^3/s
c. 3.60 ft^3/s
d. 8.25 ft^3/s

45. Determine the average velocity of a 10-in. sewer main with a flow of 1.2 ft^3/s?

a. 0.004 ft/s
b. 2.22 ft/s
c. 0.65 ft/s
d. 4.11 ft/s

46. A pump at a wet well must raise 2.3 mgd to a height of 60 ft to storage. What horsepower is needed if the pump has an efficiency of 85 percent?

a. 24.20 hp
b. 0.03 hp
c. 28.47 hp
d. 0.47 hp

47. What is the flow rate in mgd in a wet well that lifts water to a height of 75 ft if it has an efficiency of 90 percent from a 30 hp pump?

a. 2.05 mgd
b. 0.09 mgd
c. 2.38 mgd
d. 0.04 mgd

48. How many pounds of a liquid with a specific gravity of 0.92 is available if there are 4,500 gal in the tank?

    a. 540 lb
    b. 587 lb
    c. 37,530 lb
    d. 34,515 lb

49. What is the specific gravity of a liquid if it weighs approximately 40,000 lb with a volume of 5,391 gal?

    a. 0.13
    b. 0.89
    c. 7.42
    d. 1.12

50. A rectangular tank that is 10 ft long, 6 ft wide, and 3 ft deep has a depth of flow of 2 ft and drains in 2 h. What is its pumping rate in gpm?

    a. 7.48 gpm
    b. 11.22 gpm
    c. 673.20 gpm
    d. 60.00 gpm

51. How many million gallons per day flow through a channel that is 5 ft wide, flowing at a depth of 18-in. at a velocity of 2.3 ft/s?

    a. 17.88 mgd
    b. 0.99 mgd
    c. 11.15 mgd
    d. 0.28 mgd

52. If 8 mgd flows through a channel 2 ft wide with a velocity of 1.3 ft/s, what is its depth?

    a. 6.19 ft
    b. 3.08 ft
    c. 19.04 ft
    d. 4.76 ft

53. How many 10 yd^3 dump trucks are needed to remove soil from a trench with the following dimensions: 4 ft wide, 10 ft deep, 1,200 ft long?

    a. 20 trucks
    b. 45 trucks
    c. 178 trucks
    d. 1,800 trucks

54. What is the average depth of a trench with the following information: 5 ft wide, 1,000 ft long, with a total of 1,300 yd^3 of soil removed from it?

    a. 1.30 ft
    b. 7.02 ft
    c. 13.10 ft
    d. 35.00 ft

55. What is the average depth of water in a tank that has a diameter of 25 ft with a total of 35,000 gal in it?

    a. 7.51 ft
    b. 9.54 ft
    c. 19.81 ft
    d. 25.00 ft

General

1. A wastewater collection system consists of all of the following except:

    a. gravity sewers and force mains
    b. manholes
    c. clearwell
    d. pumping equipment

2. One of the reasons wastewater collection systems fail is because of which of the following?

    a. Blockages in the sewer mains, businesses, or homes
    b. Power failures experienced at the water treatment plant
    c. Consistent operation and maintenance practices
    d. Raw sewerage being conveyed into the wastewater treatment plant

3. What effect does a leaking sewer main have on the system?

    a. Improper installation and the storage of materials surrounding the leaking pipe
    b. The gradual washing away of bedding material and soil adjacent to the pipe
    c. The potential for the street above the sewer main to collapse
    d. All of the above

4. Which of the following are not the proper categories of collection system customers?

    a. Office, retail, and manufacturing
    b. Neighborhoods and processing plants
    c. Domestic, regional, and state services
    d. Residential, hotels, and resorts

5. Collection system operators are usually responsible to perform which of the following?

    a. Operate and maintain water treatment plants
    b. Operate and maintain the collection system to convey wastewater to the plant
    c. Focus on operating the system with minimal training
    d. Use power-operated equipment to only inspect system components

6. Which of the following are not a result of problems within the wastewater collection system?

    a. Inadequate flow capacities in the area served
    b. Proper bedding around the sewer mains
    c. Joint failure
    d. Pipe erosion

7. Root growth is a primary cause of pipe damage that allows which of the following?
   a. Percolation and precipitation
   b. Permeation and penetration
   c. Infiltration and exfiltration
   d. Infestation and indigestion

8. Which of the following are not disease-causing organisms or agents contained in raw sewerage?
   a. Chlorine
   b. Bacteria and protozoa
   c. Viruses
   d. Bioaerosols

9. Combined sewer overflows and sanitary sewer overflows (CSOs/SSOs) generally happen during wet weather events, lift station failures, and
   a. decreased flows into the system
   b. disinfection procedures
   c. pipe failure and/or settlement
   d. manhole stabilization

10. Which of the following is not required by the proposed Sanitary Sewer Overflow (SSO) Rule released by the USEPA in January 2001?
    a. Capacity, management, operation, and maintenance (CMOM)
    b. A survey to be completed by the customer
    c. Prohibition on SSOs
    d. Reporting, recordkeeping, and public notification

11. Which of the following is not part of the National Pollution Discharge Elimination System (NPDES) permit legislation?
    a. The control of all discharges of potential pollutants
    b. Regulation of discharges from industries and sanitary landfills
    c. The turbidity of the water at the water treatment plant intake
    d. Return irrigation flows

12. Wastewater system pumps perform which of the following?
    a. Replace the need for gravity flow mains
    b. Lift the wastewater to a higher elevation
    c. Pump sewerage without the need for maintenance
    d. Convey sewerage only where the topography is favorable

13. Which of the following is not part of the total quantity of wastewater conveyed?
    a. The amount of source water treated and distributed
    b. Quantities of commercial and industrial wastes
    c. Groundwater infiltration and surface water inflow
    d. Quantities of residential wastes

14. Which of the following is not used to convey sewerage to wastewater treatment plants?
    a. Building sewers
    b. Laterals and branch sewers
    c. Main sewers and interceptors
    d. Booster pump stations

15. A low-pressure collection system is used because of which of the following?
    a. The expenses associated with pumping sewerage
    b. The topography and ground conditions are not suitable for a conventional gravity system
    c. An increased slope in one area of the collection system
    d. Larger sized collection system piping
16. Which of the following are components of a low-pressure collection system?
    a. Gravity sewers, holding tanks, grinder pumps, and force mains
    b. Sewer laterals, manholes, interceptors, and pump stations
    c. Storm drains, interceptors, and siphons
    d. Closed laterals, booster pumps, and surge tanks
17. Which of the following are not the principal components of a vacuum collection systems?
    a. Sewer laterals, holding tanks, and drainage
    b. Gravity sewers, grinder pumps, and manholes
    c. Gravity sewers, interface unit, and pumps
    d. Gravity sewers, force mains, and pumps
18. Which of the following rigid pipe materials are not commonly used in collection systems?
    a. Reinforced concrete pipe (RCP)
    b. Vitrified clay pipe (VCP)
    c. Nonreinforced concrete pipe (NRCP)
    d. Steel pipe
19. Which of the following are not commonly used pipe joints within the collection system?
    a. Gasketed joints
    b. Flanged joints
    c. Epoxy joints
    d. Cement-mortar joints
20. Which of the following is true about collection system manholes?
    a. Manhole bases are always cast with reinforced concrete
    b. All manhole barrels are cast in place sections
    c. Manhole steps and rungs do not have to be corrosion resistant
    d. Manholes used in areas with a high groundwater or corrosion are often made with and/or lined with fiberglass
21. Which of the following are not examples of confined spaces in collection systems?
    a. Manholes, tank vessels, and pits
    b. Buffer zones, exclusion zones, and traffic zone
    c. Silos and storage bins
    d. Hoppers and vaults
22. Which of the following are examples of atmospheric hazards as defined by the Occupational Safety and Health Administration?
    a. Condensed air, moist vapor, and aerated dust
    b. Atmospheric oxygen concentrated between 19.5–21.5 percent
    c. Flammable gas, vapor, and combustible dust
    d. Mist with the volume of air less than 10 percent of its lower explosive limit

23. Whenever an operator is in a manhole, which of the following should be observed?
    a. The operator observing should secure objects to prevent them from falling into the manhole
    b. The oxygen in the manhole should be continuously monitored
    c. All precautions and safety procedures should be followed
    d. A periodic assessment of the worker's condition and the air should be performed
24. Which of the following is not an important requirement to trench excavation?
    a. The edges of all surface cuts do not have to be smooth and perpendicular when excavating
    b. Exits should be provided at 25-ft intervals at depth of trenches greater than 4 ft
    c. Depth should be deep enough to protect the pipe from freezing, surface traffic, streams, or river erosion
    d. Trench should be no more than 1 to 2 feet wider than the outside diameter of the pipe
25. Which of the following is not true about shoring?
    a. Sheeting and trench boxes are forms of shoring
    b. The length of time the excavation will be open and depth dictate the type used
    c. All types are not designed to prevent cave-ins
    d. Shoring requirements are dictated by codes and laws
26. Which of the following is not a type of shoring?
    a. Spread shoring
    b. Hydraulic shores, screw jacks, and air shores
    c. Solid sheeting
    d. Cylinder shoring and shield shoring
27. Which of the following does not apply to avoiding cave-ins in trenches 5 ft deep or greater?
    a. Shoring is required may be based on type and condition of soil
    b. The spoil bank is to be kept back a minimum of 2 ft from the edge of the trench
    c. Under ideal conditions, uprights should be spaced a maximum of 8 ft apart
    d. The ladder should extend at least 18 in. above the trench
28. Which of the following is not a part of the lock out/tag out process?
    a. Notify all affected employees regarding the use of the process
    b. Shut down the equipment or machine normally, if it is in operation
    c. Isolate the equipment or machine and restore energy to it during repairs
    d. Lock out/tag out the equipment using the assigned employee tags
29. Which of the following categories is not a hazardous material operators are exposed to within collection systems?
    a. Biodegradable dilutions
    b. Corrosives
    c. Flammables and explosives
    d. Asphyxiants and infectious agents

30. Which of the following are considered infectious agents?

    a. Cholera, cryptosporidiosis, and meningitis
    b. Hepatitis, gastroenteritis, and typhoid fever
    c. *E. coli*, Giardia, and salmonella
    d. Bacteria, protozoa, helminths, and viruses

31. Inspection and testing of existing collection systems ______________.

    a. identify existing and potential problem areas
    b. are not intended to be used as part of the new construction acceptance process
    c. assist in evaluating the severity of the detected problems
    d. provide detailed information about system problems

32. Collection system problems may be caused or created by which of the following?

    a. Proper design or construction
    b. Appropriate operation, maintenance, or use
    c. Outside forces, events, or natural disasters
    d. All of the above

33. Which of the following is not true about infiltration and inflow?

    a. They are not recognized as a major cause of operating problems
    b. They cause poor performance or failure of collection systems
    c. They often cause a hydraulic overload to the system
    d. They could result in the surcharge of manholes

34. Which of the following is not true regarding manholes inspections?

    a. Manholes are subject to the negative impact of vibration and vehicular traffic
    b. Manholes may settle at a different rate than the connected sewer mains
    c. Manholes can perform their job effectively without the attention given to other assets
    d. Manholes can reveal the content and volume of flows in the system

35. Which of the following is not true about closed-circuit television (CCTV) inspections of sewers?

    a. Positive and reliable information about the sewer's condition
    b. The accurate location of sewer problems
    c. An uncommon approach to inspecting sewer mains
    d. A great tool for determining which sections of pipe to rehab, repair, or replace

36. Closed-circuit television (CCTV) inspections do not allow system personnel to observe which of the following?

    a. The pipe's structural condition
    b. The most accurate rate of flow
    c. The integrity of the pipe joints
    d. The location of leaks (infiltration and inflow)

37. Which of the following is not true about smoke testing?

    a. It is a cost-effective way to identify inflow
    b. It usually delivers immediate results when used
    c. Equipment costs are often more expensive than other inspection processes
    d. It uses nontoxic smoke-filled air

38. Which of the following is not an accurate benefit of the use of smoke testing?
    a. Does not reduce the possibility of dangerous gases getting into a building
    b. Helps to identify the location of needed rehab/repairs
    c. Reduces the cost of testing contaminated inflow water
    d. Aids in meeting one of the US Environmental Protection Agency's mandates to eliminate infiltration and inflow
39. Smoke testing of collection systems is not used to determine which of the following?
    a. The sources of the water entering the system
    b. The identification of the types of pipe material in the system
    c. If buildings or residences are connected to the system
    d. The location of illegal connections, broken pipe sections, and lost manholes
40. Which of the following is not true about dye testing used in the collection systems?
    a. Increase the rate of sewer flow into a building
    b. Used to determine if facilities are connected to the system
    c. Can identify the cross connection between sanitary and storm systems
    d. Cost-effective way to determine if sewer facilities are connected to the system
41. Which of the following is not a use of dye testing by operators?
    a. In buildings and areas suspected of being an inflow source
    b. Should be used exclusively instead of closed-circuit television to inspect sewer mains
    c. Where a storm drain in a yard is suspected of being connected to the system
    d. To estimate the approximate velocity of the flow
42. Which of the following is not true about sanitary sewer stoppages?
    a. A stoppage can result in sewerage overflowing onto the street or in homes
    b. Stoppages occur when the pipe becomes plugged and flow is reduced
    c. All stoppages result in the flow eventually discontinuing
    d. Stoppages can surcharge a manhole and flow into an adjacent creek
43. Which of the following is not the cause of a type of stoppage?
    a. Water flowing into the system above 95°F
    b. Broken pipe and failed joints
    c. Grease and debris
    d. Roots
44. The sewer cleaning and maintenance method used is not dependent upon which of the following?
    a. The characteristics of the wastewater
    b. The fluctuation of flows
    c. The alignment/grade and physical condition of the pipe
    d. The energy used to pump the sewerage
45. All of the following are true about hydraulic or mechanical equipment used to clean sewers except which of the following?
    a. They cannot remove grit, grease, and debris from the sewer
    b. They clear emergency stoppages
    c. They perform preventive maintenance activities
    d. They use water under pressure to produce high velocities

46. The type of hydraulic equipment used to clean sewer mains include which of the following?

   a. High velocity cleaners
   b. Poly pigs
   c. Bags and kites
   d. All of the above

47. Which of the following is not considered part of mechanical cleaning equipment?

   a. Power rodders and hand tools
   b. Cutting tools, porcupines, and swabs
   c. Cheaters, extensions, and slings
   d. Bucket machines

48. Which of the following methods are more effective for releasing an emergency stoppage?

   a. Bacteria, bucket machines, and can scooters
   b. Chemicals, kites, and poly pigs
   c. High velocity cleaners, hand rodders, and power rodders
   d. Bags, poly pigs, and scooters

49. Which of the following is more effective for grease removal?

   a. Bags and chemicals
   b. High velocity cleaners, power rodders, and bacteria
   c. Bucket machines and kites
   d. Scooters and hand rodders

50. Which of the following is more effective against roots?

   a. Chemicals, high velocity cleaners, and power rodders
   b. Bags, kites, and poly pigs
   c. Bucket machines and scooters
   d. Bacteria and hand rodders

51. Which of the following is not typically a cause of grease problems in the collection system?

   a. High velocities in sewers near restaurants
   b. Low velocities in sewers from a disposal unit
   c. When high velocity in a sewer suddenly slows down
   d. A restaurant with a malfunctioning grease trap

52. Which of the following problems is a cause of sand, grit, and debris issues?

   a. Eggshells, coffee grounds, and bones from residential units
   b. Broken dishes, bones, and debris from restaurant disposal units
   c. Sand, silt, and debris from broken lines and poor joints
   d. All of the above

53. H2S gas and odor problems are not caused by which of the following in the collection system?

   a. Low flow velocities and force mains
   b. The accumulation of sand and silt
   c. Offset joints, drop manholes, and snags in the line
   d. Manholes where septic tank contents were dumped

54. Which of the following is not part of the type of nozzle selection process?
    a. Pipe size, pipe type, and pipe depth
    b. The amount of sewerage flowing through the pipe
    c. The distance to the lift station
    d. The type of cleaning or stoppage
55. Which of the following is not part of the basic categories of cleaning nozzles?
    a. Basic drilled nozzles and specialty nozzles
    b. Rotational or spinning nozzles
    c. Low-speed reverse nozzles
    d. Bottom or floor cleaners
56. Which of the following types of nozzles are better for cleaning grease and clearing a light amount of roots?
    a. Basic drill nozzles
    b. Rotational or spinning nozzles
    c. Bottom or floor cleaning nozzles
    d. Specialty nozzles
57. Which of the following perform a better cleaning on force mains and other large diameter sewers while in service?
    a. Power rodders
    b. Chemicals and high velocity cleaners
    c. Bags, kites, and poly pigs
    d. Bucket machines
58. Which of the following is not true about power rodders?
    a. They pull or push cleaning
    b. They are equipped as continuous or sectional rods
    c. They are commonly used on grease deposits, debris and root removal
    d. They are not used for clearing stoppages
59. Which of the following is not associated with continuous rodding machines?
    a. They can be changed out without replacing the whole reel
    b. They are quickly and easily loaded compared to sectional machines
    c. They cover a greater distance than sectional machines
    d. The rods must be cleaned as they pass through the rollers
60. Sectional rodding machines are not known by which of the following?
    a. A broken rod can usually be repaired in the field
    b. The rods are pushed through a drive head
    c. Couplings on the sectional rods seldom cause wear on the guide hose liner
    d. A broken sectional rod is easier to retrieve than one from a continuous rodder
61. Which of the following is used as a finishing tool to scour pipe after the cleaning operation?
    a. Porcupine
    b. Auger
    c. Sand leader
    d. Swivel

62. Which of the following is the best tool to relieve stoppages in pipe over 6 in. in diameter?
    a. Auger
    b. Spear heads
    c. Square bar corkscrew
    d. Spring blade cutter

63. Which of the following is the best tool to follow the auger or root saw to pull back roots/debris?
    a. Double point corkscrew
    b. Spring blade cutter
    c. Spear heads
    d. Porcupine

64. Which of the following is not true regarding the procedures for recovering broken rods from a sectional rod machine?
    a. Measure the amount of the rod pulled back
    b. Couple and install the pickup tool to the section
    c. Couple the rods on either side of the broken section
    d. The process is more difficult than that for the continuous rod machine

65. Which of the following is not true for recovering a broken section of a continuous rod machine?
    a. The coupling tool stays connected when a section of the continuous rod machine breaks
    b. May require fashioning a tool to the end to retrieve the lost section
    c. The pickup tool is designed to attach itself to the lost section
    d. With the pickup tool attached, the push rod can usually be located with a slow and steady rotation

66. Which of the following is not true as a benefit of using hand rods?
    a. Can be used when and where mechanized methods are not effective
    b. Provide a quick solution for many minor stoppages
    c. Often used in easements where power rodders are not as effective
    d. Are not used for emergencies at night and on weekends

67. Which of the following is not a hand rodding tool or accessory?
    a. Round wire corkscrew
    b. Push in tool
    c. Rod guide holder
    d. Rod wheel safety stand

68. Which of the following is not a true consideration for clearing a stoppage with hand rods?
    a. Rod the shortest distance to the stoppage
    b. It is easier to rod into an overflowing upstream manhole
    c. If rodding upstream, the cleared stoppage will flow toward you
    d. If rodding downstream, place a trap downstream to catch debris

69. For routine cleaning with hand tools, which of the following apply?
    a. The proper tool will be dependent on the sewer's condition
    b. Never use the square bar auger
    c. The cleaning and rodding tool cannot be pulled out by hand from the opposite manhole
    d. Remove debris from the manhole before disassembling the guide tool and brace

70. Which of the following is not a way solid debris enter the collection system?
    a. Through loose or misaligned joints and connections
    b. Manhole lids
    c. Service connections and discharges into the system
    d. In a pressurized well-maintained force main

71. Which of the following is not true about power bucket machines?
    a. Used where hydraulic cleaning techniques are not the best option
    b. Better at removing larger, heavier objects from sewers
    c. Not a better option than high velocity cleaners in failing mains
    d. More effective than high velocity cleaners in large trunk and interceptor mains

72. A power bucket machines does not work more effectively than high-velocity machines in which of the following situations?
    a. As a waterless cleaning method
    b. For removing large amounts of heavy dirt and debris
    c. Before and after a repair to remove debris
    d. In smaller-diameter pipes

73. Which of the following is not a power bucket machine accessory?
    a. Roller and yoke assembly
    b. 3-point jack roller
    c. Bucket hooks
    d. Slack jack

74. Which of the following are objectives of cleaning equipment maintenance?
    a. To prolong equipment life and maintain efficiency
    b. To keep the equipment in minimal use and repair
    c. To delay or postpone equipment failure on the job
    d. All of the above

75. Which of the following is not an indication of needed maintenance on a high-velocity cleaner?
    a. Sluggish reel return
    b. Partially full hold tank at the end of a run
    c. Odd or suspicious sounds from the intake valve seats
    d. Higher or lower RPMs than normal

76. Which of the following is not part of the maintenance for a sectional rodder?
    a. Inspect the belt tension and wear
    b. Check the wear of the drive dogs
    c. The tension of the drive head chain must be loose
    d. Check the rod couplings

77. Which of the following is true for the maintenance program for a continuous rodder?
    a. Check the drive rollers and rods for signs of slippage
    b. Check the gears for the drive head, sharp or pointed teeth are the better
    c. Check the rod guide bushings for wear which do not shave the rod
    d. Replace pads for the rod clearing tool annually

78. Which of the following is/are part of the maintenance for a power bucket machine?
    a. Tighten all cable clamps weekly
    b. Only lube spindles and lower rollers when they are not in contact with water
    c. Look for loose rivets and badly worn bails
    d. Only straighten stabilizer legs after they have been bent for some time

79. The proper application of chemicals in sewer mains cannot be used to control which of the following?
    a. Grease and odors
    b. Insects and rodents
    c. Corrosion and roots
    d. Leaking and poor joints

80. Which of the following is true about selecting and using chemicals in the collection system?
    a. Most herbicides, insecticides, fungicides, and rodenticides require registration with the US Environmental Protection Agency (USEPA)
    b. All chemicals will immediately clear a stoppage
    c. Laboratory or desktop demonstrations of chemicals will accurately reflect what is needed
    d. All chemicals used require a USEPA registration number

81. Which of the following is not true about root control?
    a. The entry point of roots into the sewer is usually above the flow line
    b. Cutting or pruning sewer roots prevents thicker, faster, and stronger regrowth
    c. Roots that grow from the top of the pipes are called "veils"
    d. Roots thrive in the sewer atmosphere

82. When selecting and using chemicals in sewer mains, which of the following should be provided by the manufacturer/provider?
    a. The methods of application and optimal weather conditions
    b. The concentration of chemicals needed and the required contact time
    c. The cost and any safety precautions
    d. All of the above

83. The effective use of foaming in sewer mains ________________.
    a. destroys existing roots and prevents regrowth
    b. does not contain a fumigant and a growth inhibitor
    c. attaches itself to the surface of the pipe
    d. is best when smoke testing was last conducted

84. Which of the following ways is the best for applying foam to the roots in the sewer system?
    a. Applying the foam using the basic nozzle
    b. Spraying a layer or filling a section with foam
    c. Flooding a manhole and letting the foam leak into the sewer main
    d. Applying the foam to the section of pipe below the flow line
85. Which of the following is true regarding flooding sewers with root control chemicals?
    a. Flooding sewers and building service laterals can only be done in high-velocity sewers
    b. The method is commonly used in sewer mains ranging from 4–6 in. in diameter
    c. Only the downstream end needs to be isolated with the flooding process
    d. In sewer mains ranging from 6–12 in. in diameter, flooding is used to control roots
86. The rate and volume of grease buildup in the sewer main is dependent on which of the following?
    a. The amount of grease carried by the wastewater
    b. The flow and velocity of the wastewater
    c. The size of the sewer
    d. All of the above
87. Which of the following statements is untrue regarding grease accumulation in sewer mains?
    a. Were two sewer mains intersect is a potential location for grease buildup
    b. Smaller sized mains experience grease accumulation more than larger ones
    c. Grease traps and interceptors are required if grease is in the discharge
    d. Pumping the liquid out of the trap or interceptor removes problems
88. The critical components of an effective rodent and insect control program are which of the following?
    a. Sample, test, and notify
    b. Detect, test, and respond
    c. Identify, prevent, and treat
    d. Identify, treat, and notify
89. Effective insect control may not be achieved by which of the following?
    a. Performing the effort as a separate process from other maintenance activities
    b. Coordinating with local health departments
    c. Periodically flushing slow or stagnant lines during warm-weather months
    d. Reducing potential environments for insects to accumulate
90. Hydrogen sulfide causes which of the following problems?
    a. Possible paralysis of the respiratory system of operators
    b. Loss of capacity in the sewers
    c. Corrosion and possible collapse of sewers, structures, and equipment
    d. All of the above

91. Hydrogen sulfide-related odor control issues are not caused by which of the following?
    a. High-velocity flows
    b. Long transmission lines in the collection system
    c. High temperatures
    d. Poorly maintained collection systems
92. Which of the following components is/are not part of a comprehensive odor control program?
    a. Investigate odor complaints in a timely manner
    b. Routinely perform sewer maintenance
    c. Infrequent monitoring of the sewer hydrogen sulfide concentration
    d. Effective sewer construction, rehab, and repair
93. Which of the following statements is not true regarding maintaining collection systems?
    a. All sewers should be cleaned at least once every five years
    b. Manholes should not be sealed in areas where odors continuously vent
    c. Trap manholes should be inspected and cleaned quarterly
    d. Siphons should be cleaned to reduce the accumulation of debris and the generation of hydrogen sulfide gas
94. Scrubbers provide which of the following benefits to collection systems?
    a. Increasing the air pressure in the system to move the odors along
    b. Controlling odors in many situations
    c. Adsorbing odorous compounds into the carbon media
    d. b and c
95. The degradation of the wastewater in the collection system has which of the following impacts on the wastewater treatment plants?
    a. High slime levels and lower flows are good for the wastewater treatment plants
    b. The wastewater arriving negatively impacts the treatment and generates odors
    c. The strength of the sewerage does not negatively impact the water returning discharges
    d. The degradation produces desirable effects at the wastewater treatment plant
96. Wastewater system repairs provide which of the following benefits?
    a. Correct damage and deterioration
    b. Remove obstructions from the system
    c. Correct bad designs and poor construction activities
    d. All of the above
97. Which of the following is not an example of an improper or defective tap?
    a. A protruding tap
    b. A misaligned tap
    c. A wet tap
    d. A hammer tap

98. Which of the following is not true regarding sewer pipe bedding?

    a. Usually consists of sand, gravel, crushed concrete, or stone
    b. Most utilities do not have specific requirements regarding bedding
    c. Provides support around pipe walls
    d. Offers a firm base the entire length of the pipe

99. Which of the following is not true about placing pipe segments?

    a. Smaller diameter pipe can be placed directly on the bedding
    b. Larger diameter pipe should be supported using slings and lifting hooks when placed
    c. The bell end of the pipe should be facing downstream
    d. The impact force and the weight of the large diameter pipe should never be used to achieve the proper grade

100. Which of the following is a part of an effective backfill effort?

    a. Protecting the pipe from movement
    b. Properly placing the material around and over the sewer pipe
    c. Avoiding the breaking of the pipe on impact into the trench
    d. All of the above

101. Which of the following is not the ideal sewer trench backfill material?

    a. Material free of any object that will harm the pipe
    b. Non-native granular material
    c. Easily compacted
    d. Immediately available

102. Restoring the surfaces does not involve which of the following?

    a. Excess soil and rock excavated from the trench
    b. The use of control density fill material
    c. Backfilling and compacting using mechanical or hand equipment
    d. Restoring the ground surface to its original condition or better

103. Which of the following is not true about bypass piping?

    a. Used where a temporary plug is needed to facilitate repair
    b. Used to isolate pipe sections that experience high or inconsistent flows
    c. Used instead of rehabilitation or internal inspections
    d. Used where repairs may result in backups or overflows

104. Main line repairs are required in which of the following situations?

    a. Where sewer lines repeatedly show partial or complete stoppage
    b. Broken and collapsed pipe
    c. Pipe with deep sags or severely offset joints
    d. All of the above

105. Manhole repairs consists of which of the following?

    a. Not requiring the bottom flow channel to be smooth to reduce friction
    b. Making sure the frame and cover are secure
    c. The top being slightly above the grade of the surrounding surface
    d. Barrel and cone sections having minimal cracks and deterioration

106. Which of the following are not components of an outside drop manhole?

   a. A reverse wye or tee
   b. A wooden check dam
   c. Select bedding material
   d. Drop connection pipe and fittings

107. Which of the following is not true of pick holes and vent holes in manhole covers?

   a. This is how surface runoff, flood waters, and ponded stormwater enter the system
   b. Vent holes are used for a gradual release of gases and odors
   c. Vent holes are plugged to increase the amount of flow into the system
   d. All of the above

108. Which of the following does not mean that service lines for buildings, houses, and taps must be repaired?

   a. Lines are structurally deficient
   b. There is a problem at the tap
   c. Lines are not able to properly convey flows
   d. There is no cleanout in place

109. Which of the following is not a type of a trenchless repair method for sewer pipe?

   a. Slip lining
   b. Insitu flow
   c. Spray lining
   d. Pipe bursting

110. Which of the following are part of a precast concrete manhole?

   a. Base, invert, and bench
   b. Trunk section and rigid fixed rings
   c. Barrel and cone sections
   d. Frame and cover

111. Which of the following is not true regarding air testing sewer mains?

   a. Air can leak through cracks to find vapor leaks
   b. Small cracks eventually become larger ones that can release wastewater into the environment
   c. High-pressure air is used to test sewers
   d. Cracks often attract roots and cause infiltration

112. Which of the following statements is not true regarding mandrel testing of sewers?

   a. Not recommended to be used for rigid, nonflexible pipe materials
   b. Usually performed after backfill and compaction but before surface restoration
   c. Used to measure for deflection and offset joints
   d. Will identify when the building sewer connection is protruding into the mainline

113. Which of the following is true regarding lift stations?

   a. Lift stations convey wastewater or stormwater from a higher elevation to a lower one
   b. Lifting the wastewater is done using hydraulic pumps
   c. Lift stations represent a major capital expenditure for any agency or community
   d. Failures of lift stations and force mains are minor with insignificant impact on the area

114. Collection system operators play a vital role in maintaining lift stations by

   a. preventing catastrophic failures affecting the community
   b. requiring a minimum scope of training on lift stations
   c. not having to have a high level of knowledge and skill
   d. intermittently using maintenance programs

115. The design and location of lift stations are based on all of the following except

   a. economic factors analyzed by an engineer
   b. constructed at low points in the collection system
   c. excavation costs to maintain gravity flows and velocities become excessive
   d. groundwater table is deep and restricts installing deep mains

116. Which of the following is not true about lift stations?

   a. They are designed to move wastewater with the maximum energy consumption
   b. They are selected to provide a flow as continuous or consistent as possible
   c. They should be designed to minimize surges in the downstream sewers
   d. Those selected should require minimum maintenance

117. All of the following are a type of lift station except

   a. a standard manhole equipped with a submersible pump
   b. a manhole equipped with a dry-suction pump
   c. a factory prefabricated package plant
   d. a station designed to pump large volumes of wastewater

118. Which of the following is not used as a level sensor in a manhole used as a lift station?

   a. Pneumatic bubbler system
   b. Multiple floats
   c. Exposed-electrode controllers
   d. Level pressure transducer

119. Of the following choices, which do not apply to dry well stations?

   a. They isolate pumps, motors, and electrical controls in a dry well
   b. Usually provides a cleaner, safer environment for operation and maintenance
   c. These stations do not vary in size and are not prefabricated
   d. They are a common type of lift station used in collection systems

120. Which of the following is frequently overlooked during lift station design?

   a. Properly sizing the pump
   b. Determining the adequate size of the force main
   c. Evaluating the most appropriate flow through the force main
   d. The addition of air relief and air vacuum valves

121. The operating levels of a lift station are determined by
   a. the high level is limited to the invert of the incoming gravity pipe
   b. the lowest level is defined by the center line of the pump impeller
   c. both the high and low water levels in the wet well
   d. all of the above

122. Programmable logic controllers
   a. are not found in most supervisory control systems
   b. are used in upgraded or retrofitted supervisory control systems
   c. use modules to manually control pump sequencing
   d. are not usually programmed using a laptop

123. Which of the following is not true about motors used in lift stations?
   a. Electric motors are the most common drivers for pumps of lift stations
   b. Motors are one of the key components of a lift station
   c. Direct current and alternating current are the motor classifications
   d. Direct current pump motors are the primary ones used

124. Which of the following are components of a typical, non-clog, vertical, raw wastewater centrifugal pump?
   a. Mid-bearing, grease flange, jack stop, and grinder sleeve
   b. Flexible coupling, shaft washer, pump base, and adjuster
   c. Impeller, volute, wear rings, and packing rings
   d. Floor line, pump frame, suction base, and lateral ring

125. Which of the following do not apply to mechanical seals?
   a. Leakage of raw sewerage from the top of the stuffing box is eliminated
   b. Create more wear on the pump shaft or shaft sleeve
   c. Lead to less power consumption
   d. Will last for years when installed and operated properly

126. Motor and control terms are more likely
   a. ampere, bearings, insulation, thermistor, and windings
   b. adjusted temperature, directed delay, switch gear, and dampener
   c. current adjuster, insertion tube, rotor flow, and thermaflex
   d. control expander, modular coupling, stator flux, and gain controller

127. The types of gate valves used in the wastewater collection system are
   a. bonnet and wedge valves
   b. cone and pressure-reducing valves
   c. non-rising stem and outside screw and yoke
   d. flat disc and packing valves

128. All of the following are hazards at lift stations except
   a. insufficient O_2, high presence of explosive and toxic gases
   b. adequate drainage
   c. exposed electrical or medium/high voltage equipment or repairs being conducted
   d. unsafe stairs and walkways

129. Which of the following best determines the frequency of visits to the lift station?

   a. The number of customers in the community
   b. The type of wastewater being stored
   c. The potential damage to the station and the surrounding area
   d. All of the above

130. Failures at lift stations are not a result of

   a. pumps on line and controlling wet well levels
   b. failure of the level sensor or other control equipment
   c. plugged impeller or lines
   d. stuck or blocked check valve

131. Problems at lift stations may be associated with which of the following categories?

   a. Suction, discharge, strength of sewerage, and the size of the lift station
   b. Power, control systems, pumping systems, and structures
   c. Corrosion, distribution of waste, distance to the plant, and topography
   d. Ambient temperatures, pump sizes, surrounding environment, and pipe sizes

132. Other than using bleach where possible, controlling hydrogen sulfide in force mains involves which of the following?

   a. Dosing air into the force main upstream of the lift station
   b. Dosing gaseous pure oxygen into the force main upstream of the lift station
   c. Installing ventilation equipment at the force main outlet
   d. Dumping caustic downstream of the lift station

133. Which of the following is not a type of alternating current?

   a. Quad phase
   b. Single phase
   c. Two phase
   d. Three phase or poly-phase

134. Identify the statement that is false regarding grounding.

   a. Grounding must be taken into account wherever electrical current flows
   b. Effective grounding means the path to ground is continuous and permanent
   c. Grounding must have ample current-carrying capacity
   d. Proper grounding considers equipment and system components only

135. Short cycling or excessive starts are problems associated with the controls of which type of motors?

   a. Static drive motors
   b. Induction motors
   c. Dual phase motors
   d. Dc wound motors

136. Shaft misalignments are associated with all of the following except

   a. shaft spacing misalignment
   b. angular misalignment
   c. parallel and angular misalignment
   d. parallel misalignment

137. Cavitation is associated with all of the following except

 a. cavitation causes a reduction in the pump's efficiency
 b. cavitation can cause damage to the impeller
 c. cavitation is a result of unusually high pressures in the pump
 d. cavitation happens when a pump operates at flow rates considerably higher than designed

138. Screw pumps lift wastewater to a higher elevation by

 a. being driven by a conical gear speed driver
 b. the support of a single bearing
 c. using pressurized discharge
 d. operating at a constant speed within a housing or trough

139. Pneumatic ejectors

 a. consist of a single check valve and an ejector pump
 b. receive flows from a pressurized system and discharges them into a gravity one
 c. are used to handle limited or low flows with relatively large solids
 d. do not contain a float device or ball

140. A pneumatic ejector system is

 a. a centrifugal system
 b. a tankless system
 c. a forced-air system
 d. an open-air system

141. Piston pumps

 a. vary based on the manufacturer
 b. cylinders do not vary with pressure and gallons per minutes
 c. are commonly used in low-velocity cleaners
 d. use a fluid cylinder, piston, intake and discharge valves

142. The capacity of a piston pump is determined by all of the following except:

 a. the smaller stroke allows more water to enter
 b. the number of strokes per minute
 c. the length of the strokes
 d. the diameter of the piston or plunger

143. Closed-coupled pumps

 a. have the motor and pump as two units
 b. have an extended shaft
 c. mount the volute surrounding the impeller on the top of the motor
 d. all of the above

144. One of the advantages of a closed-coupled pump is

 a. its size permits installation in a tight space
 b. it is used exclusively in cold water systems
 c. the mating flanges do not have to be clean
 d. it is not supported by the motor bearing

145. Which of the following best describes submersible pumps?
    a. They are very different from the closed-coupled pumps
    b. The pump requires extensive maintenance
    c. The motor used is waterproof and mechanical seals keep water out
    d. The pump does not use the temperature of the liquid pump to cool the motor

146. Centrifugal or trash pumps
    a. are not designed to be used as portable pumps
    b. do not have to be primed to be used efficiently
    c. can be hindered by high suction lift
    d. all of the above

147. The positive displacement diaphragm pump
    a. uses a rigid membrane in the horizontal cylinder
    b. uses the concentric movement of the rod to alternate from suction to pressure
    c. is a piston-type system
    d. uses the common ball check valves

148. The types of impellers used in centrifugal pumps are
    a. closed flow, coupled flow, volute-driven, and radius flow
    b. mixed flow, propeller, radial flow, and turbine
    c. transient flow, modified suction, improper flow, and induction flow
    d. conical flow, max flow, positive suction, and shaft-driven

149. The most common impellers used in wastewater collection system centrifugal pumps
    a. cannot consistently pass solids with less changes of plugging
    b. are closed impellers for small pumps
    c. are closed-type radial or semi-closed for large pumps
    d. are not required to have a specific clearance between the tip of the impeller and volute

150. If packing is not properly maintained
    a. water or wastewater can potentially contaminate the bearings
    b. air will not be allowed to enter the pump
    c. the shaft and shaft sleeve will be well maintained
    d. an increase in suction will occur

151. Which of the following is not an indication of packing failure?
    a. Packing is not well lubricated through the gland
    b. Top gland ring is in poor condition and the bottom gland ring is in ok condition
    c. Rings are extruding past the gland follower
    d. Minimal leakage at start up

152. The type of bearings not commonly used in pumps in wastewater operations are
    a. ball bearings
    b. sleeve bearings
    c. stroke bearings
    d. roller bearings

153. Couplings are used to connect the source of power to a shaft using

a. electrical, diesel, gas, or hydraulic pressure
b. air processed through a condenser
c. converting energy from wastewater
d. indirect current converted to energy

154. Compressors are commonly used in wastewater operations and maintenance by

a. decreasing the pressure of air and gas
b. activating ejectors and pump control systems
c. not having a suction line or filter
d. using a complex diaphragm or a simple vane type

155. Proper lubrication of equipment and tools

a. is not one of the most important maintenance functions
b. is not one of the easiest maintenance jobs
c. is to increase the friction between surfaces to generate energy
d. is intended to change the friction between surfaces using the lubricant

156. One of the primary reasons to rehabilitate a sewer in the past was

a. to decrease the flow in a particular area
b. to restore the structural integrity of a failed line
c. to increase the number of surcharged sections
d. to balance the level of deterioration with flows

157. The first step in defining a rehabilitation program is to

a. develop a comprehensive written description of collection system components
b. analyze the condition of collection system assets
c. set goals for the rehabilitation effort
d. prioritize collection system assets

158. Sewer system evaluations are intended to:

a. increase the hydraulic loads on the collection system
b. balance the increase of combined sewer overflows and sanitary sewer overflows
c. reduce the infiltration/inflow (I/I) problems in existing sewers
d. analyze I/I issues only

159. High levels of infiltration and inflow can

a. have minimal to no impact on the operation of the system
b. increase the system's capacity to handle peak flows
c. increase the system's ability to maintain the capacity reserved for future growth
d. have serious negative impacts on the operation of the system

160. Flow monitoring is a primary tool for determining high infiltration and inflow by

a. measuring flow variations instantaneously
b. identifying and measuring base flow, infiltration, and inflow
c. disregarding peak flow conditions as anomalies
d. evaluating water consumption records without adjusting for inaccuracies

161. Exfiltration results in

    a. the potential to contaminate the groundwater in the area
    b. being measured differently than infiltration and inflow is captured
    c. a measurable increase in expected flows
    d. not being harmful to any adjacent public drinking water supply

162. The major source of rainfall induced infiltration and inflow are

    a. gate valves, flow monitoring, and pipe rehabilitation
    b. invert alignments, sewer construction, and bad inspection practices
    c. house connections, manholes, and sewer mains
    d. improper fittings, pressure surges, and sanitary sewer overflows

163. The structural integrity of the sewer pipe is negatively impacted by

    a. surface loads directly over the pipe accounted for within the design range
    b. traffic loads included in the design
    c. anticipated and present use conditions
    d. changes in the land surface over the sewer pipe

164. The flow data acquired as part of the sewer system evaluation survey (SSES) should

    a. detail a written description of all leaks
    b. be organized in a format that indicates flow variations by the time of day
    c. provide historical data from problem areas
    d. document discharge permit violations

165. Which of the following is not a course of action for a rehabilitation program?

    a. Stepping up the preventive maintenance program to extend the life of the assets
    b. Completely replacing sections of sewer in immediate danger of failure
    c. Prepare to replace the worst sections when they fail
    d. Gradually replacing sections of sewer pipe that are deteriorating

166. The process for setting priorities for sewer rehabilitation primarily include

    a. the flow within the sewer pipe
    b. the distance from the wastewater treatment plant
    c. the depth of the sewer main
    d. the consequence of failure

167. Which of the following is not a problem common to all construction?

    a. The selection of pipe materials
    b. Traffic control and disruptions
    c. Excavation dewatering
    d. Shoring requirements

168. Chemical grouting is the most widely used method

    a. for sealing leaking pipe joints and circumferential cracks
    b. which incorporates the use of cement-mortar and epoxy
    c. that only uses heat to cure the elements
    d. that produces a permeable layer around the pipe

169. The following are all considered robot drilling tools except

a. diamond finger drill
b. diamond pear drill
c. diamond drill
d. diamond slotted drill

170. The primary advantage of trenchless rehabilitation over open-cut construction is:

a. when soil or groundwater conditions are normal and not a problem
b. in rural areas with little to no underground interference
c. in downtown areas where there is a high possibility of business disruption
d. the increased cost of trenchless over the cost of open-cut construction

171. Cured-in-place pipe is

a. a conventional form of pipe replacement
b. a resin-saturated, flexible fabric
c. a non-structural pipe rehabilitation solution
d. pushed into place and cured using grout

172. Sliplining, also called pipe insertion, is

a. one of the newest forms of continuous structural linings
b. used by pushing a new rigid liner pipe in place
c. identified by a smaller pipe being inserted into a larger pipe
d. employed where there are obstructions in the pipe

173. High-density polyethylene (HDPE) is a solid-walled pipe that

a. requires extensive excavation as part of its process
b. the diameter has no bearing on the trench size
c. reduces the cross-sectional area of the pipeline
d. does not require the annular space to be grouted

174. Liners are a repair method that

a. is not used to control infiltration and inflow (I/I)
b. reconditions the sewer system with new, smooth pipe
c. does not require an excavation
d. depends on whether the procedure costs more than Insituform

175. Manhole defects are a result of

a. corrosion or deterioration of the wall and rungs
b. an extremely small amount of infiltration and inflow (I/I)
c. being properly grouted between sections
d. the frame being properly sealed to the top of the adjusting ring

176. Manhole coatings

a. are successful even with improperly prepared surface prep
b. can be effective without bypassing the wastewater
c. can be best on moist surfaces of concrete
d. are used to repair the deteriorated interior of manholes

177. The major causes of accidents or injuries are
   a. hand injuries, leg injuries, and air-deprived areas
   b. falls, sprains, strains, and being struck by objects
   c. hearing injuries, electrical injuries, and confined space injuries
   d. head injuries, defects in personal protective equipment, and uneven surfaces

178. The top administrator of the wastewater utility should
   a. promote safety programs throughout the organization
   b. approve safety-related policies and programs
   c. ensure adequate budgets and resources are allocated for safety
   d. all of the above

179. The duties of coordinating safety activities with regard to insurance agencies is the job of
   a. safety department staff
   b. the top administrator
   c. the manager
   d. the field supervisor

180. Monthly safety meeting topics should
   a. be less comprehensive than tailgate meetings
   b. cover subjects more specific to the work environment
   c. be presented by safety personnel only
   d. last for more than a few hours

181. Workplace spaces are considered "confined" if
   a. there is a reduced exposure to potential contaminants
   b. the entrapment hazards have been eliminated
   c. their configuration hinders the worker's access into and out of the space
   d. there is adequate air flow

182. The employer should provide confined space training and education
   a. only when the worker's duties change
   b. for any employee required to work in a permit-required space
   c. to make sure the worker passes their certification exams
   d. to evaluate the employee's lack of knowledge on the subject

183. Wastewater collection system facilities should be protected by
   a. requiring all visitors to abide by all employee rules
   b. encouraging public parking at critical buildings to demonstrate transparency
   c. being aware of the hazard chemicals used and how to neutralize them
   d. storing emergency numbers in an isolated location

184. The high-risk level of a potential terrorist attack is associated with which color?
   a. Red
   b. Orange
   c. Yellow
   d. Blue

185. In the event of a work-related injury,

    a. the impacted employee should notify their supervisor within 24 hours
    b. the supervisor is required to investigate the injury within 7 days
    c. the safety committee should review the employee's information and the supervisor's report
    d. the incident should only be documented based on a majority vote of the committee

186. The disciplinary policies of the agency are intended to

    a. bring unacceptable behavior to the attention of the employee
    b. establish punitive discipline for safety violations
    c. disregard consequences to better develop employees
    d. create an environment where employees are discouraged to make mistakes

187. Occupational Safety and Health standards contains the regulations most applicable to work in the collection system in

    a. Subpart B, Part 1910
    b. Subpart E, Part 1910
    c. Subpart C, Part 1900
    d. Subpart A, Part 1910

188. Subpart I, Part 1910, Occupational Safety and Health standards covers

    a. medical and first aid
    b. fire protection
    c. personal protective equipment
    d. materials handling and storage

189. An effective administration should

    a. not be concerned about being respected by employees
    b. allow employees to manage all of the activities of the agency
    c. discourage workers to challenge decisions
    d. ensure quality operations and maintenance programs to maximize the success of the organization

190. A detailed written plan for the operation and maintenance of the system should

    a. be equipped with specific punitive disciplinary policies
    b. focus on the manager's desires
    c. be developed only by the workers responsible for operation and maintenance
    d. address the agency's mission and goals

191. Which of the following is a reasonable goal for a wastewater utility?

    a. Respond to customer's concern when not performing other duties
    b. Restrict the public's access to all facilities during normal working hours
    c. Properly convey wastewater to treatment facilities with minimal disruptions
    d. Maintain critical equipment on an as-needed basis

192. Objectives and task/procedures for the operation of the collection system

    a. should not be established to align with previous goals
    b. may often conflict with the overall goals of the organization
    c. should be clear and concise enough to be reasonably achieved
    d. all of the above

193. The procedures for the application, selection, orientation, and training of workers should

   a. be subjective and based on the history of the organization
   b. consider what is needed to effective manage the assets of the organization
   c. be oriented toward employee concerns only
   d. be more reflective of the individual and not standardized

194. The superintendent along with the assistance of the Human Resources (HR) department should

   a. establish standards for the effective selection of personnel
   b. allow the HR department to conduct the entire hiring process
   c. only select the type of employees the superintendent desires
   d. not be concerned about the employees selected with the goal of training them

195. The selection process should consider

   a. reviewing the applications and resumes prior to the interview
   b. establishing standards for what is acceptable for employees
   c. using the interview to further evaluate the candidates
   d. all of the above

196. Operator certification is required by most states

   a. without regard for the impact on the public
   b. to recognize the importance of the knowledge of effective operations and maintenance practices
   c. considering the skills taught by the utility only
   d. using the knowledge and practices of the best utility in the area

197. Tool and equipment requirements consider all of the following except

   a. the location of utility staff
   b. the size of the collection system
   c. the workload of the staff
   d. the age and condition of the assets

198. High velocity cleaners, trailer units, and truck-mounted units

   a. are low to no maintenance vehicles
   b. operate effectively regardless of the capacity of the water carried
   c. are straight hydraulic jetting units and combination units
   d. are primarily intended for catch basins and storm drains

199. Closed-circuit television units

   a. are only available and effective in one size
   b. are selected based on the various needs of the organization
   c. that are trailer-mounted are not advantageous because they have to be towed
   d. that are truck-mounted are not well-suited for high traffic areas

200. A good preventive maintenance program for equipment

   a. is based on when and how each piece of equipment is to be maintained
   b. will maximize the amount of equipment needing repairs
   c. will perform repairs when the equipment is at or near failure
   d. is accomplished with minimal regard for downtime

201. Equipment should be replaced
   a. when the model owned is no longer supported by the manufacturer
   b. when it cost less to maintain the unit than to purchase a new one
   c. when it ceases to perform its required work
   d. when the records for the unit are not consistently maintained

202. Wastewater collection system facilities
   a. consist of shops, yards, and storage facilities only
   b. are often located far from the wastewater treatment plants, particularly small systems
   c. should be separated without regard to function
   d. should be considered for de-centralization based on the least-cost analysis

203. Maps of the wastewater collection system
   a. prevent maintenance operators from consistently locating assets
   b. provide an accurate record of underground system components
   c. are flexible enough to remain the same even if the system has expanded
   d. are usually not needed to perform maintenance

204. For the sake of maintenance, collection system maps should reveal the location of which of the following assets?
   a. Water mains, water valves, and fire hydrants
   b. Other underground utilities and critical landmarks
   c. Building sewers, manholes, and sewer mains
   d. All of the above

205. A typical software package for wastewater collection systems
   a. will not track reactive maintenance activities
   b. compiles inventories of mains, building sewers, and manholes only
   c. can control cost to operate and control the collection system
   d. may not be considered essential for the operation and maintenance of a small system

206. Computerized maintenance management systems (CMMSs)
   a. are the most effective tool for large utilities only
   b. can store a limited amount of data about the system
   c. allows for information to be rapidly manipulated to be accessible and properly used
   d. is a component that makes the utility vulnerable due to liabilities

207. Operation and maintenance recordkeeping primarily incorporates:
   a. good information from bad data input into the system
   b. tracking computerized maintenance management system failures and breakdowns
   c. several databases that collaborate to effectively manage assets
   d. offers information to be used by the public only

208. A geographic information system (GIS)

a. uses anecdotes as simple pieces of information regarding assets
b. is a program that combines mapping with detailed information about assets
c. captures information automatically and allows retrieval manually only
d. automatically updates information without input from operators

209. The supervisory control and data acquisition (SCADA) system provides computer monitoring for

a. maps, line segments, and quadrant locations
b. pipe types and invert elevations
c. alarm, control, data acquisition, and response
d. systems using RTAs and PLDs

210. Wastewater collection system records must be kept

a. to comply with state and federal regulations
b. as long as necessary or until there is no place to house them
c. by paper to reduce the possibility of losing them by computer
d. and consistently backed up in the system they were generated on

211. A wastewater collection system report

a. is enhanced by incomplete field notes and inaccuracies
b. should be submitted without a review to maintain its accuracy
c. written on a laptop or notebook is guaranteed to be accurate
d. documents the work performed by field crews and work teams

212. Public relations involve

a. informing the public on a need to know basis regarding utility information
b. providing a service to the public with courtesy, sincerity, and transparency
c. focusing more on the work to be done and engaging the public during emergencies only
d. communicating with the public only to avoid liability

213. The structure of the wastewater utility

a. is mandatory to separate the authority of the organization from other departments
b. focuses primarily on the division of labor and not equipment
c. should identify the function of each element and its relationship to operations and maintenance
d. has little to no impact on the efficiency of field crews and work teams

214. Power and authority should be delegated to others

a. to identify who will make a good supervisor
b. to allow decisions to be made at the lowest practical level of the organization
c. to determine who to blame if something goes wrong or is not done
d. to identify how many people a supervisor can effectively manage

215. An organization with an effective system of accountability will
    a. allow employees to perform their tasks without instruction to evaluate their knowledge
    b. promote high safety risk employees to reduce the possibility of accidents or injuries
    c. use it to discipline undesirable employees
    d. establish work standards as well as ways to measure performance

216. The proper management of a wastewater utility
    a. requires the manager, or supervisor to plan, organize, and staff to efficiently control activities
    b. allows for flexibility in all areas of operation of the utility
    c. can be accomplished with minimal long-term goals
    d. should reflect the personality of the manger to be successful

217. The appropriate size and type of field crews and work teams should be determined by
    a. the employees that work best together
    b. the scope of activities, program goals, objectives, and geographic locations
    c. the problems in a specific area of the system
    d. specialization, regardless the size of the utility

218. A wastewater utility's ability to handle emergencies
    a. is a measure of how well the management staff is trained
    b. should be important, but not the first priority
    c. shows the dysfunction in the preventive maintenance effort
    d. is demonstrated by their success with removing main/service line blockages and correcting lift station failures

219. Which of the following priorities is a greater liability risk?
    a. A flooded home or business
    b. An overflowing manhole
    c. A missing manhole cover
    d. An overflowing cleanout

220. Flooding at a home, business, or at a street caused by a mainline blockage
    a. should be responded to by giving the impacted customers the assurance that the utility is responsible
    b. should be addressed by staff following approved agency policies
    c. should be attended to after the claims or legal department has given approval
    d. should leave the damaged property in place until all of the necessary pictures are taken

221. An emphasis on preventive maintenance in the collection system
    a. tends to reduce the number of stoppages and often restores full flow capacity
    b. has little to no positive impact on the service life of the system
    c. works well whether the staff is specifically trained for it or not
    d. benefits the agency regardless of their knowledge of the location of assets or not

222. Training field crews
   a. provides a platform to demonstrate how to efficiently and safely perform assigned tasks
   b. has a minimal positive impact on the effectiveness of experienced crews
   c. can often impede their ability to finish tasks in a timely manner
   d. prevents them from really knowing how to improve

223. Crews making repairs
   a. should be based on priority and the need for an immediate repair
   b. have authority above respecting private property when working in easements
   c. should properly barricade an open excavation at the end of each day
   d. all of the above

224. Lift station maintenance
   a. has changed over the years as a result of a decrease in related regulations
   b. improves the reliability of the asset and results in fewer component/system failures
   c. is less expensive than a gravity system and costs less to maintain
   d. is not as important for a small utility as it is for a large one

225. Industrial waste inspections (pretreatment facility inspections)
   a. allow the customer to have their samples tested by a lab apart from the utility
   b. are taken without any regard to charges associated with usage
   c. are made to verify compliance with the agency's sewer-use and industrial waste ordinances
   d. provide information without making recommendations for compliance or violations

226. Maintenance programs for collection systems
   a. are not critical because the systems were designed to be self-cleaning
   b. are not the responsibility of the supervisor to manage or direct
   c. are better when reactive instead of trying to predict or prevent failures
   d. are a result of conditions developed that interrupt flow and diminish the design capabilities

227. Which of the following is not considered a Level 1 Maintenance task?
   a. Recording pump operating levels
   b. Analyzing pump discharge pressures
   c. Performing supervisory system maintenance
   d. Observing incoming line voltage and current

228. Performance indicators for collection system are based on
   a. the number of stoppages per mile of sewer main
   b. the number of failures per lift station
   c. the cost of maintenance of the sewer main and lift station
   d. all of the above

229. Standard permit conditions addressing capacity, management, operations and maintenance (CMOM)

 a. were intended to replace National Pollutant Discharge Elimination System (NPDES) permit requirements for sanitary sewer systems
 b. were signed without allowing the incoming administration time to review
 c. were intended to minimize sanitary sewer overflows and combined sewer overflows
 d. were proposed to protect humans and not aquatic life

230. Factors that contribute to poor system performance and deterioration are

 a. construction oversight, proper inspections, and management of project
 b. inadequate flow capacities, pipe corrosion, and unexpected population growth
 c. the management of wastewater treatment plant hydraulic overloads
 d. the appropriate monitoring of infiltration, inflow, and exfiltration

231. The proper financial management of a utility involves

 a. financial stability for the agency
 b. careful budgeting
 c. providing capital improvement funds for future utility expansion
 d. all of the above

232. A utility in good financial standing is an agency with

 a. an operating ratio as well as a coverage ratio above 1.0
 b. the total revenues received from its customers less than total expenses
 c. the average revenues without potential variations not considering expenses
 d. the total of user fees, hook up charges, and special income paying the debt service only

233. The Governmental Accounting Standards Board (GASB) Standard 34

 a. discourages the reporting of depreciated assets
 b. is intended to make financial reporting for state and local governments more comprehensive and easier to complete
 c. allows agencies to use an asset management system to avoid having to use a modified approach
 d. requires a critical assessment of high capital assets

234. The effective operation and maintenance of wastewater collection systems

 a. is minimally impacted by the materials selected and how they are installed
 b. considers what it has the capacity to purchase currently
 c. is to keep the system performing as designed and intended
 d. can be performed consistently without performance measures

This page intentionally blank.

II. Wastewater Treatment, Questions

Math

1. Convert 1,000 gpm to million gallons per day.

   a. 86,400,000 or 86.4 mgd
   b. 1,440,000 or 1.44 mgd
   c. 24,000 or 2.4×10^{-2} mgd
   d. 48,000 or 4.8×10^{-2} mgd

2. Convert 1 m^3/s to million gallons per day.

   a. 0.65 mgd
   b. 21.6 mgd
   c. 22.8 mgd
   d. 15.3 mgd

3. A flow of 7.2 mgd needs a polymer dose of 12 mg/L or ppm. The polymer is a solid that is a 15 percent solution by weight (density 1.0 g/cm^3). Determine the feed rate.

   a. 45 gal/h
   b. 42 gal/h
   c. 24 gal/h
   d. 32 gal/h

4. The concentration of mixed liquor suspended solids (MLSS) in the aeration basin should be 3,500 mg/L. Currently, it is 4,200 mg/L. The plant aeration tank's volume is 69.0×10^4 ft^3. Determine the volume of liquid to be removed from the aeration tank to achieve the 3,500 mg/L MLSS. The density of the aeration tank with the MLSS is 1.05 g/cm^3. Neglect MLSS in the incoming waste stream.

   a. 200 m^3 or 52,834 gal
   b. 19.54 m^3 or 5,072 gal
   c. 13.67 m^3 or 3,614 gal
   d. 1 percent of volume under aeration

5. Determine how much sludge should be wasted to remove 35,500 lb from a system, ignoring the solids coming into the influent of the aeration basin, given the following:

   The sludge removed from the clarifier has a concentration of 2 percent solids by weight.
   The aeration basin is 3,000,000 gal with a clarifier following.
   The overall mixed liquor suspended solids (MLSS) level in the aeration basin is 4,200 mg/L.
   The weight of MLSS to be removed is 35,500 lb.
   The clarifier underflow concentrates the sludge to 2 percent solids.
   The density of the clarifier underflow is 1.05 gm/cm^3.

   a. 1,360,000 gal (5,148.2 m^3)
   b. 223,322 gal (845.4 m^3)
   c. 18,000 gal (68.14 m^3)
   d. 680,000 gal (2,574.1 m^3)

6. You take a sample 1 L of the aeration basin. In the laboratory, you find that the tare weight of the container is 140 g. The tare weight of the filter is 3.52 g. After filtering and drying a 1,000 cm^3 sample of aeration tank mixed liquor, the final weight of the container and the sample is 154.23 g. After ignition, the weight of the container plus sample is 149 g. Calculate the mixed liquor suspended solids (MLSS) and the mixed liquor volatile suspended solids (MLVSS) concentrations.

   a. MLSS 5,400 mg/L, MLVSS 3,750 mg/L
   b. MLSS 6,930 mg/L, MLVSS 3,410 mg/L
   c. MLSS 1,523 mg/L, MLVSS 1,473 mg/L
   d. MLSS 3,650 mg/L, MLVSS 2,111 mg/L

7. River flows are routinely measured in either ft^3/s, m^3/s, or in some cases m^3/d. Wastewater discharges in the United States are measured in million gallons per day. The significance of this calculation is that wastewater discharges are diluted by river or stream flows in a ratio basis. Convert 1,300 ft^3/s to m^3/d, m^3/s, and million gallons per day.

   a. 2.4×10^6 m^3/d, 27.7 m^3/s, 847.6 mgd
   b. 3.18×10^6 m^3/d, 36.81 m^3/s, 840.1 mgd
   c. 4.8×10^6 m^3/d, 55.53 m^3/s, 1,268 mgd
   d. 2.8×10^6 m^3/d, 33.3 m^3/s, 760.8 mgd

8. A Parshall flume measures influent flow from a pipe with a 15-in. (38.1-cm) diameter. The flow is measured at 0.83 ft in a Parshall flume with a 6-in. throat. Assuming that the measurements are correct and the flume is calibrated, the equation of flow is Q (ft^3/s) = $2.08 \times H^{1.58}$, where H is in feet. Compute the flow in million gallons per day.

   a. 1.55 mgd
   b. 2.08 mgd
   c. 1.01 mgd
   d. 0.481 mgd

9. A digester has a conical bottom. The digester is a 38-ft diameter with a 14-ft sidewall. The bottom cone of the digester is 8 ft below the sidewall. Find the volume in cubic feet.

   a. 16,000 ft^3
   b. 15,900 ft^3
   c. 18,900 ft^3
   d. 15,877 ft^3

10. A plant has a flow of 6.5 mgd. The aeration basin is 435,000 ft^3. The two clarifiers have 82-ft diameters with a 12-ft side water depth. What is the hydraulic retention time in the aeration basin? What is the retention time in the clarifiers? Do not assume recycle.

   a. 24 h aeration basin, 3.5 h clarifiers
   b. 12.1 h aeration basin, 3.5 h clarifiers
   c. 36 h aeration basin, 7.0 h clarifiers
   d. 14 h aeration basin, 6.5 h clarifiers

11. A tank that is 8 ft in diameter and 20 ft tall is being filled with a hose from a pumped flow. The pumping rate is unknown, but the depth of liquid in the tank can be measured. When the tank is half full, the liquid level rises at 1 ft per 20 min. How long will it take for the tank to become 80 percent full, and what is the estimated pumping rate?

   a. 75 min and 32.4 gpm
   b. 90 min and 25.2 gpm
   c. 180 min and 12.5 gpm
   d. 120 min and 18.8 gpm

12. The influent biochemical oxygen demand (BOD) is about 250 mg/L. The effluent BOD on the permit is 10 mg/L. Determine the BOD removal rate for the plant.

   a. 81 percent
   b. 91 percent
   c. 96 percent
   d. 99 percent

13. The bottom valve in a vertical tank is opened. The tank has a 3-in. valve (d = 0.25 ft), and the tank is 12 ft tall and has a 6-ft diameter (D = 6 ft). The drainpipe is at the bottom of the tank, or $h_f = 0$. The average time required to drain the tank is given by the following formula:

   Time (T) = $0.4155 \times (D^2)/d^2 \times (\sqrt{h_o} - \sqrt{h_f})$
   where T = time required to drain tank (in seconds), D = diameter of tank, d = diameter of pipe, h_o = initial height of liquid in the tank, h_f = final height of liquid in the tank. Find the approximate time required to drain the tank.

   a. 54.35 min
   b. 12.16 min
   c. 13.82 min
   d. 18.5 min

14. Most wastewater is chemically balanced for nutrients required to encourage bacterial growth. Chemical oxygen demand (COD) is most often used to indicate the amount of carbon in wastewater. The ratio of COD to total Kjelldahl nitrogen to total phosphorus (COD:TKN:P) allows one to determine the chemical balance of the wastewater for conditions ideal to promote bacterial growth. This is often abbreviated as C:N:P.
   In wastewater that ratio is approximately 125:20:1. Given the following wastewater characteristics, which one is the limiting factor for good growth?

   COD 300 mg/L, TKN 59 mg/L, P 2 mg/L

   a. Phosphorous
   b. Nitrogen
   c. Carbon
   d. Other (not shown)

15. Determine the food-to-microorganism ratio (F:M) in the aeration basin of an activated sludge plant.

    Flow in = 5 mgd
    Waste strength = 250 mg/L five-day biochemical oxygen demand (BOD5)
    Aeration basin volume = 335,000 ft^3
    Mixed liquor suspended solids (MLSS) = 2,950 mg/L

    a. 0.33
    b. 0.25
    c. 0.20
    d. 0.17

16. A chlorine contact tank is required to have a 30-min detention time and a dose of 5 mg/L at all times. The average plant influent flow rate is 1 mgd, and the daily peak flow is 2.5 times the average. The chlorine contact tank is 7,000 ft^3. Find the detention time in the chlorine contact basin at peak flow.

    a. 28 min
    b. 30 min
    c. 61 min
    d. 15 min

17. Calculate the quantity of 11 percent sodium hypochlorite (NaOCl) solution required for disinfection at average and peak flow rates in gallons per hour, given the following:

    The molecular weight of NaOCl is 74.45
    NaOCl has 1.06 g of chlorine (Cl^2) per liter of 100 percent NaOCl
    At 11 percent strength, 1 g of chlorine will be provided by 9.1 L of NaOCl solution
    The chlorine dosage required for disinfection is 5 mg/L
    The flow is 1 mgd at average flow, and 2.4 times that for peak flow

    a. 216 gph normal and 518 gph peak flow
    b. 126 gph normal and 302 gph peak flow
    c. 70 gph normal and 168 gph peak flow
    d. 110 gph normal and 264 gph peak flow

18. Given the following discharge values for a wastewater treatment plant over a two-week period, compute the average flow and standard deviation of the data. Assuming a normal distribution, what upper and lower values should contain about 95 percent of the flow data?

Date>>>	**1**	**2**	**3**	**4**	**5**	**6**	**7**
Flow	22.99	26.61	21.99	25.53	21.84	19.36	19.27
Date>>>	**8**	**9**	**10**	**11**	**12**	**13**	**14**
Flow	20.86	30.4	21.66	27.01	31.1	25.5	29.13

    a. Upper 28.43, Lower 20.56
    b. Upper 24.52, Lower 16.90
    c. Upper 36.38, Lower 11.87
    d. Upper 30.40, Lower 20.11

19. What is the effect of a 1-ft change in the water level in the main aeration tank?

    a. 0.43 psi
    b. 2.31 psi
    c. 0.5 psi
    d. 1.0 psi

20. At sea level, what is the maximum theoretical height to which water can be drawn into a pump?

    a. 15.6 ft
    b. 26.3 ft
    c. 33.9 ft
    d. 45.1 ft

21. What is the sludge age in a plant, ignoring the clarifiers, given the following information?

    Total reactor volume = 500,000 gal
    Mixed liquor suspended solids (MLSS) = 3,500 mg/L
    Wasting rate = 15,000 gpd
    Waste activated sludge (WAS) concentration = 4,500 mg/L

    a. Approximately 15 days
    b. Approximately 40 days
    c. Approximately 30 days
    d. Approximately 26 days

22. Calculate the dose required to remove 4 mg/L of phosphorus in a 2.5 mgd plant, given the following:
    Ferric chloride ($FeCl^3$) molecular weight (MW): 162.5: weight metal: phosphorous (P): 1.8:1
    Specific gravity 1.37 at 35 percent
    Gallons of solution required to remove 1 lb of P: 1.32

    a. 520 gpd
    b. 90 gpd
    c. 110 gpd
    d. 50 gpd

23. The sludge sent to the anaerobic digester produced 10 ft^3 of gas per pound of volatile solids (VS) destroyed. How much gas is produced per day given the following:

    Sludge pumped = 10,000 gal at 9.1 lb/gal/day
    Solids concentration = 6 percent
    Initial VS concentration = 65 percent
    Final VS concentration = 12 percent
    Digested sludge solids concentration = 12 percent

    a. 310,069 ft^3/d
    b. 20,356 ft^3/d
    c. 28,938 ft^3/d
    d. 54,600 ft^3/d

24. What is the approximate weight of the sludge generated when thermally drying sludge from a centrifuge given the following:

    The initial feed is 12,000 gpd at 12 percent solids at 9.5 lb/gal
    After drying the sludge, moisture content is 10 percent

    a. 13,680 lb
    b. 15,048 lb
    c. 23,761 lb
    d. 11,403 lb

25. If the flow through a 75-ft-diameter clarifier is 3.5 mgd, what is the surface overflow rate?

 a. 400 gpd/ft^2
 b. 654 gpd/ft^2
 c. 792 gpd/ft^2
 d. 1,013 gpd/ft^2

26. If a clarifier launder ring is limited to 10,000 gpd/ft, how many feet of launder ring will be required for two 150-ft-diameter clarifiers when the peak flow is 3,500,000 gpd?

 a. 280 ft
 b. 350 ft
 c. 420 ft
 d. 110 ft

27. Calculate the chlorine dose and the demand of a gas chlorinator, given the following.

 The gas chlorinator has the capacity to feed 75 lb of chlorine (Cl^2) per day
 It is only feeding 25 lb/d
 The wastewater treatment plant has a flow of 1.32 mgd
 When the chlorine test is performed after 30 min, the residual is 0.5 mg/L

 a. 1.91 mg/L dose = 1.41 mg/L demand
 b. 3.34 mg/L dose = 2.84 mg/L demand
 c. 0.59 mg/L dose = 0.09 mg/L demand
 d. 2.25 mg/L dose = 1.95 mg/L demand

28. A 5-L sample of wastewater is aerated for 30 min, and the aeration is turned off. After the 30-min aeration, the dissolved oxygen (DO) concentration is measured at 3 mg/L. After the aeration was stopped, the DO level drops to 1.5 mg/L after 15 min and to zero (0.0) mg/L after 25 min. What is the oxygen uptake (consumption) rate of the wastewater after 25 minutes?

 a. 25 mg/L/h
 b. 5 mg/L/h
 c. 7.2 mg/L/h
 d. 8 mg/L/h

General

1. The valve pictured here is a ________________.

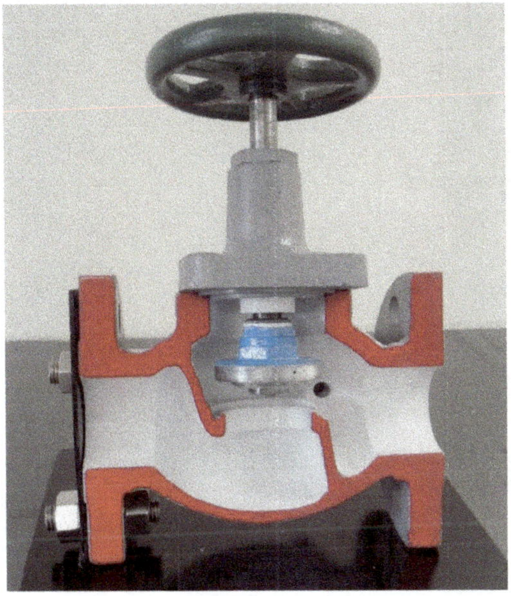

Figure source: *Bayport Valve*

a. gate valve
b. globe valve
c. ball valve
d. plug valve

2. The advantage of the valve pictured here is ________________.

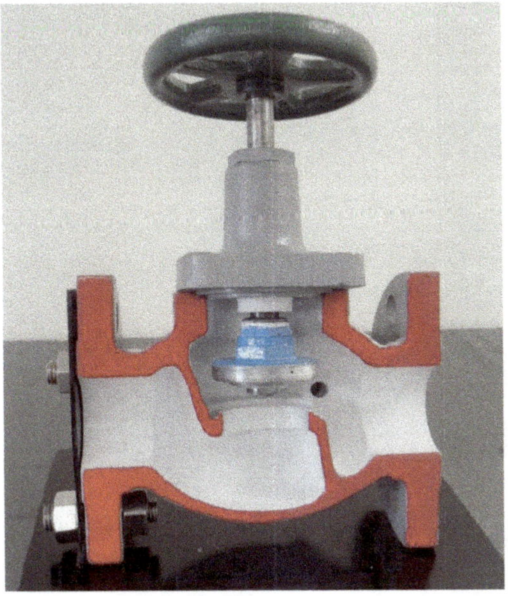

Figure source: *Bayport Valve*

a. good flow control
b. low head losses
c. good resistance to plugging
d. all of the above

3. The valve pictured here is a ________________.

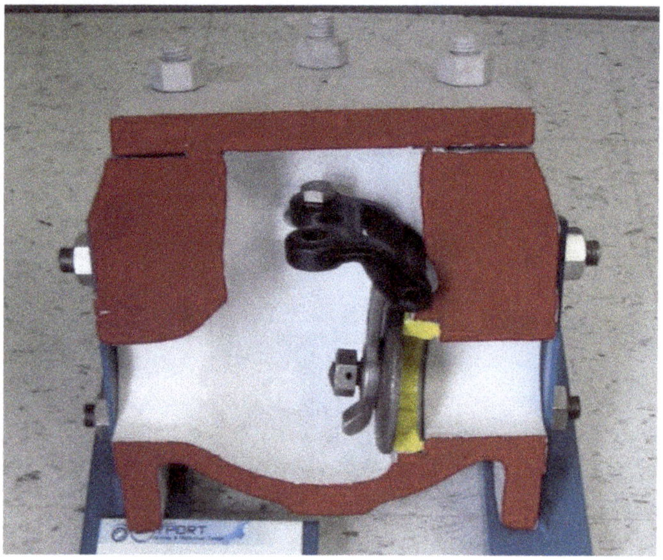

Figure source: *Bayport Valve*

a. ball valve
b. globe valve
c. plug valve
d. check valve

4. The purpose of the valve pictured here is to

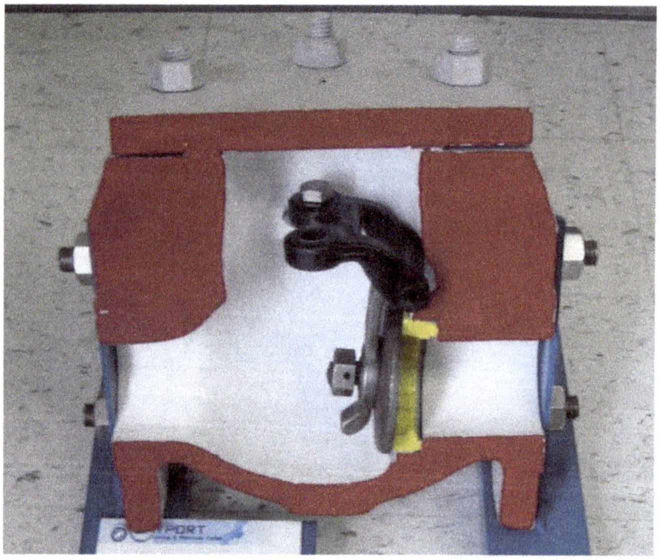

Figure source: *Bayport Valve*

a. ensure that flow goes in one direction only.
b. serve as a flow regulation valve.
c. maintain pressure on the system.
d. prevent water hammer.

5. The valve pictured here is a ________________.

Figure source: *Bayport Valve*

a. ball valve
b. check valve
c. plug valve
d. gate valve

6. The valve pictured here shows a type of ________________.

Figure source: *Bayport Valve*

a. pressure relief valve
b. safety valve
c. plug valve
d. gate valve

7. The valve pictured here is a ________________.

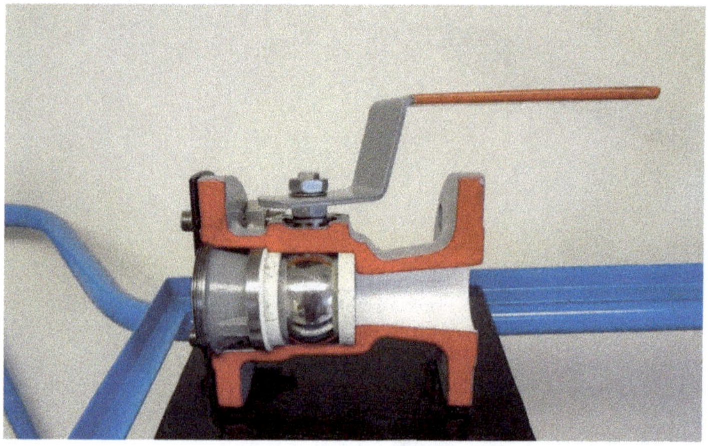

Figure source: *Bayport Valve*

a. plug valve
b. ball valve
c. gate valve
d. globe valve

8. The valve pictured here is a ________________.

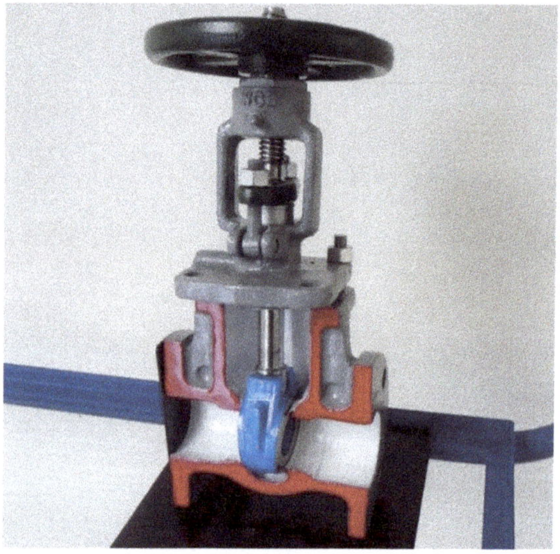

Figure source: *Bayport Valve*

a. rising stem gate valve
b. gate valve
c. globe valve
d. plug valve

9. What are the advantages of rising stem valves?

a. One can determine whether the valve is open or closed by inspection.
b. The valve provides a very secure seal.
c. The repairs are simpler.
d. The valve is inherently safer because of its body and design.

10. What are the advantages of plug and ball valves in liquid service?

   a. Good flow control and easy determination of valve position.
   b. The valve seat/face connection is wiped when the valve is opened or closed, and the valve will cut stringy or long solids and provide good closure.
   c. The valve is suitable for higher solids contents.
   d. Because the valves are ¼ turn shutoff, it is easy to tell when they are closed or open.

11. What is the problem with gate valves in wastewater service?

   a. The valves can collect grit and fail to seal properly.
   b. The head losses through the valve are quite variable.
   c. Many of the valves leak at the valve stems.
   d. The valves need frequent maintenance.

12. The valve pictured here is a ________________.

Figure source: *Bayport Valve*

   a. gate valve
   b. plug valve
   c. butterfly valve
   d. check valve

13. What is the problem/challenge with the valve pictured here related to sewage service?

Figure source: *Bayport Valve*

   a. It's difficult to obtain a good valve closure as a result of solids accumulating on the valve sealing perimeter.
   b. It requires a high amount of service/maintenance requirements.
   c. Stringy and large solids can accumulate on the valve face.
   d. It has high torque and marginal flow regulation, making it difficult to operate.

14. Why should a valve not be opened or closed quickly on larger pipes and for longer lines if it creates water hammer?

    a. The water hammer is annoying but generally harmless.
    b. The starting or stopping of the flow can cause solids to deposit on both sides of the valve.
    c. A pressure surge can destroy any pumps connected to the system.
    d. The developed pressure surge can rupture the pipe.

15. What information must be collected and recorded when collecting a sample of wastewater and transmitting it to the laboratory?

    a. Date, time started and stopped, sample ID number
    b. Temperature of liquid, chain of custody form
    c. Instructions to the laboratory and which analyses to perform
    d. All of the above

16. The recommended material and preservation technique for most samples is

    a. to freeze the sample and analyze it within 3 months, storing it in glass.
    b. to cool the sample to 4°C and analyze it within 48 h, storing it in glass or plastic.
    c. no special requirement.
    d. to collect it in polyethylene with a polypropylene cap or in glass.

17. Regarding automatic samplers being used to collect grab and composite samples, the following statements may be true or false. Select the best answer.

    a. The flow proportioning sampler should always be connected to a flowmeter.
    b. Flow proportion samples collect composite samples in one container. Time proportional samplers collect samples in individual containers.
    c. Periodic grab sampling is adequate if the sample volumes are adjusted with the flows.
    d. All of the above are true.

18. In a sample of wastewater from an industrial source that has been collected and analyzed, the values reported by the laboratory are just over the permit limits for five-day biochemical oxygen demand (BOD5), and total suspended solids (TSS) are just over the facility permit limit. Given that there are inaccuracies in the analytical process, what should be done before contacting the company about their permit exceedances?

    a. Resample the source one more time.
    b. Set up a monitoring program to confirm the discharge.
    c. Set up a monitoring program with split sampling (one set of samples analyzed by the utility and one by the company) to verify the results.
    d. Nothing—refer the company to the utility's enforcement group.

19. Before using an automatic sampler, what needs to be checked?

    a. Thoroughly clean and rinse all tubing in contact with liquid, thoroughly clean the sample container, and replace the tubing if necessary.
    b. Thoroughly clean the sampling container and check the tubing for plugging.
    c. Disassemble and service the sampling pump to ensure correct operation.
    d. Use new tubing and sampling containers only.

20. Which of the following statements referring to sampling error is accurate?

    a. Simple composite sampling is the most inaccurate, but there is not much difference between sampling methods.
    b. Most sampling methods consistently under collect the true discharges by 1 to 10 percent.
    c. Multiple samples always need to be collected.
    d. If the flow is uniform and does not change, one sample is sufficient.

21. What is the maximum holding time for wastewater samples used for biochemical oxygen demand (BOD) analyses?

    a. 48 hours at 4°C
    b. 72 hours at 4°C
    c. Two weeks if the sample is preserved
    d. One week for most organics if they are preserved at pH < 2

22. To determine the phosphorus balance in a wastewater treatment plant, from where should samples be collected?

    a. Influent, effluent, and waste activated sludge
    b. The aeration tank and the primary clarifier
    c. Sludge sent to digester
    d. Chlorine contact tank

23. The test for five-day biochemical oxygen demand (BOD5) measures oxygen depletion in a sample over five days. What are the constraints of the test?

    a. The final dissolved oxygen must be above 0.2 mg/L and must contain a nitrification inhibitor.
    b. The seed must be acclimated to the waste, and the analysis must include a correction for the seed
    c. The temperature must be maintained at 20°C ± 0.5°C for five days.
    d. All of the above.

24. What is the physical difference between a gas and a liquid?

    a. Water is a liquid, and air is a gas.
    b. A liquid is incompressible while a gas is compressible.
    c. Water at high temperatures (steam) is a gas and is indistinguishable from air.
    d. Both are incompressible.

25. If the optimal level for dissolved oxygen in a utility's aeration tanks is 2.0 mg/L, and the system is running at 1 mg/L, what should be done?

    a. Solids should be wasted to reduce the mixed liquor suspended solids and oxygen demand, and the diffusers should be checked for plugging.
    b. Additional blowers should be added.
    c. The influent waste streams should be checked for higher than normal loadings.
    d. The diffusers should be changed out from coarse bubble to fine bubble diffusers.

26. What is a grit chamber used for?

    a. Reducing the incoming biochemical oxygen demand.
    b. Removing the grit (sand and gravel) to protect pumps and equipment.
    c. Recovering grit so that it does not get into the aeration basins.
    d. Preventing plugging of the aeration equipment.

27. What is a combined sewer?

    a. A storm sewer that also conveys sanitary flows.
    b. An overly large sewer.
    c. A sewer that conveys surface water and street drainage.
    d. A sewer that can have very highly variable flows during wet weather.

28. What is the difference between anaerobic and anoxic conditions?

    a. There is no difference.
    b. Anoxic is without free oxygen (ORP +50 to –50 mv), whereas anaerobic is without any oxygen (–100 to –225 mv).
    c. Anoxic conditions refer to a lack of oxygen, whereas anaerobic conditions refer to a type of specific reducing respiration.
    d. Anoxic is when denitrification occurs, whereas anaerobic is when hydrogen sulfide is released.

29. How are anoxic and anaerobic environments different for nitrogen and phosphorus removal?

    a. An anoxic environment has no free oxygen but has bound oxygen such as nitrates and phosphates, whereas an anaerobic environment has no oxygen.
    b. An anaerobic environment has no oxygen but is an oxidizing environment that strips the oxygen off the nitrate and phosphate molecules.
    c. The end-product of an anaerobic environment is particulate phosphorus dioxide (PO_2) and nitrogen dioxide (NO_2) as gas.
    d. An anoxic environment is an oxidizing environment, whereas an anaerobic environment is a reducing environment.

30. Which control analyses should be used to regulate the aeration supply to the various parts of the plant?

    a. Five-day biochemical oxygen demand (BOD5), 20-day BOD (BOD20), ammonia (NH_3), and nitrate (NO_3^-)
    b. Chemical oxygen demand (COD), NH_3, NO_3^-, total suspended solids (TSS), mixed liquor volatile suspended solids (MLVSS), and oxygen uptake rate (OUR)
    c. BOD5, NO_3^-, TSS, mixed liquor suspended solids (MLSS), and OUR
    d. COD, TSS, MLVSS, and OUR

31. What is a clarifier?

    a. A final step in treatment just prior to chlorination.
    b. A way to recycle the aeration solids.
    c. A device to remove suspended solid material from water by sedimentation.
    d. A device to thicken sludge.

32. What is the purpose of the rakes in sludge thickener?

    a. Remove excess water from sludge and concentrate it by stirring.
    b. Make the sludge easier to handle by breaking up lumps.
    c. Pretreat the sludge before filtration to reduce volume.
    d. Clean the bottom and sides of the clarifier.

33. What do microorganisms need to grow?

    a. Source of soluble food, nutrients, and oxygen or an electron acceptor.
    b. High levels of mixed liquor suspended solids.
    c. Particulate matter to which they can attach.
    d. Additional nutrients such as phosphorus and nitrogen.

34. What is the influence of colder temperature on treatment plant operations and efficiency?

    a. Colder temperatures slow bacterial growth and make the plant more efficient.
    b. The rate of bacterial activity decreases by about half with a 10°C (18°F) change in temperature.
    c. Colder temperatures allow better oxygen transfer rates, so the overall efficiency increases in winter.
    d. Maintenance issues increase because of ice freezing.

35. What is the optimal pH range for biological growth?

    a. 3–8
    b. 6–9
    c. 7–12
    d. >7

36. What causes extreme foaming on the surface of the aeration tank?

    a. The presence of alkyl benzene sulfonate (old style detergent) in the water.
    b. The presence of linear alkyl sulfonates (modern detergent) in the water.
    c. A local laundry discharged to the sewer.
    d. Occasional foaming is a natural occurrence.

37. How do excess quantities of fats, oils, and grease (FOG) affect the treatment plant?

    a. Accumulates on the surface of the primary and sometimes the secondary clarifier.
    b. Increases in apparent oxygen demand.
    c. Coats and accumulates inside of pipes, create plugs, and interferes with lift stations operations.
    d. Creates a layer on the top of the aeration chamber that has to be removed.

38. What is the best way to control problems arising from fats, oils, and grease (FOG)?

    a. Effective enforcement of a grease and oil collection program.
    b. Better biological treatment.
    c. Dissolve the grease with detergents.
    d. Ban grease discharges from restaurants and food industries.

39. What is the impact of food industries and hospitals on a treatment plant?

    a. They both use phosphates and strong disinfectants, which can increase treatment plant operating costs for their removal.
    b. They both contribute large quantities of blood waste, which can overload the treatment plant.
    c. The food industry can discharge high levels of organics, which will overwhelm the aeration capacity of the plant. Hospitals can discharge anti-bacterial compounds, which can kill the bacteria.
    d. Hospitals and the food industry discharge quantities of fats, oils, and grease (FOG) to the treatment plant.

40. How do toxic wastes discharges affect the plant?

    a. Toxic wastes can kill off the bacteria in the plant and cause a discharge of contaminated water by lowering or stopping bacterial metabolism.
    b. Toxic wastes cannot stop anaerobic digestion.
    c. Toxic wastes won't affect the treatment plant but can cause fish kills if the wastes pass through the plant.
    d. All of the above.

41. What is anaerobic digestion?

    a. A form of biosolids stabilization.
    b. A process to turn wastewater sludge into usable fertilizer.
    c. A method of waste treatment for solids reduction.
    d. A method of processing waste grease to generate energy.

42. Describe the conditions that adversely affect nitrification.

    a. Very low dissolved oxygen (DO), high pH, temperature less than 10°C, low alkalinity
    b. Cold temperatures, low pH (<7)
    c. High temperatures, high ratio of food to microorganism (F:M)
    d. pH above 8.5

43. When considering the occurrence of chromium, copper, lead, arsenic, and zinc in the treatment plant, which is the most toxic?

    a. All chromium is toxic in any quantity.
    b. Zinc is not toxic in most concentrations found.
    c. Arsenic and lead are very toxic even at very low levels.
    d. All are toxic at any level.

44. What is the purpose of primary treatment?

    a. To remove solid materials that pass the grit chamber.
    b. To remove solid organic materials before aeration.
    c. To capture floatable solids and reduce some of the biodegradable load to the aeration chamber.
    d. To remove suspended solids and settleable solids.

45. What happens to the solids removed in the primary clarifier?

    a. It is dried and sent to a landfill.
    b. It is sent to an anaerobic digester for further treatment.
    c. It is sent to an aerobic digester and then sent to the aeration tank.
    d. All of the above.

46. What is the purpose of the scrapers in a circular clarifier and the dragout boards in a secondary clarifier?

    a. To keep the floatable materials moving toward a central collection point.
    b. To concentrate the settleable materials and move them to the collection point.
    c. To keep the floor of the clarifier clean.
    d. To keep the solids from building up a layer and producing gas to disrupt the settling.

47. How are circular and horizontal clarifiers different?

a. The solids in a rectangular clarifier build up in a linear fashion from the entrance to the exit, whereas they build up in rings in a circular clarifier.
b. The circular clarifiers form a sludge blanket whereas the rectangular clarifiers do not.
c. The center feed well in a circular clarifier can lead to short circuiting.
d. The flow pattern and settling pattern in a rectangular clarifier are linear.

48. What are the typical removal efficiencies for primary clarifiers?

a. Organics 30–50 percent, suspended solids up to 60 percent
b. Organics 20–33 percent, suspended solids up to 40 percent
c. Organics 20–30 percent, suspended solids up to 30 percent
d. Organics 30–50 percent, suspended solids up to 75 percent

49. What is a launder ring (scum baffle) on a primary clarifier?

a. A baffle to collect the floatable materials and prevent them from going into the main aeration chamber.
b. A trap for grease and oils to prevent them from going into the aeration chamber.
c. A mechanism for keeping the floatable materials from entering the feed well of the clarifier.
d. A ring to help treat the scum in the clarifier.

50. What is the purpose of the effluent weir on a clarifier?

a. Measure the effluent flow.
b. Ensure even distribution of the flow across the clarifier surface.
c. Maintain a given level in the clarifier.
d. Prevent floatable solids from exiting the clarifier.

51. What is the function of a bar screen versus a comminutor?

a. The bar screen traps large solids for removal, whereas the comminutor shreds large particles into much smaller particles.
b. The bar screen removes fine particles equivalent to that of a comminutor.
c. A comminutor requires more maintenance than a bar screen.
d. There is no difference.

52. How are the solids in the underflow of the secondary clarifier handled?

a. All of the solids collected in the final clarifier are sent back to the head end of the plant to maintain biomass.
b. Some of the excess solids are wasted and sent through a thickening /dewatering/ digestion system for ultimate soil amendment or fertilizer use.
c. All of the solids are sent to a sludge thickener and then to an anaerobic digester where the solids are combined with some of the primary sludge and digested to destroy pathogens.
d. All solids are sent to the sludge drying beds as is.

53. What is the purpose of the return activated sludge (RAS) line?

a. It increases removals of organics, including biochemical oxygen demand (BOD).
b. It maintains the mixed liquor suspended solids (MLSS) level in the aeration tank.
c. It adjusts the solids under aeration.
d. In some systems, it allows the MLSS a second pass at the wastes.

54. What is the purpose of a diffuser?
    a. It provides air to the aeration tank.
    b. It provides mixing to the aeration tank.
    c. It creates small bubbles to increase the gas transfer rates.
    d. It helps put nitrogen into the mixed liquor suspended solids (MLSS).
55. What is a scum beach on a clarifier?
    a. A receptacle for collecting the floating solids from the surface of a clarifier.
    b. A place to collect and dry out scum on the clarifier.
    c. A modern clarifier has no need for a scum beach.
    d. All of the above.
56. What is the purpose of the center well on a clarifier?
    a. To introduce the flow into the clarifier.
    b. To redirect the flow in the clarifier toward the bottom of the clarifier.
    c. To get the piping out of the way of the rotating mechanical rakes and motor drive.
    d. To feed the flow into the center of a circular clarifier.
57. What are the advantages of coarse bubble versus fine bubble aeration?
    a. Coarse bubble aeration has better mass transfer.
    b. Coarse bubble allows higher mixed liquor suspended solids (MLSS) in the aeration tank because of better mixing than fine bubbles.
    c. Fine bubble aeration provides as much mixing as coarse bubble aeration.
    d. Fine bubble aeration has significantly better oxygen transfer than coarse bubble aeration.
58. What is the effect of fats, oils, and grease (FOG) on an aeration basin?
    a. Aside from being unsightly, FOG can cause rings around the aeration basin.
    b. FOG can greatly add to the biochemical oxygen demand (BOD)/chemical oxygen demand (COD) load of the plant.
    c. FOG can encourage the growth of filamentous bacteria.
    d. FOG requires bacteria to develop enzymes to solubilize it before digestion.
59. What is the primary cause of filamentous organisms in activated sludge?
    a. Low dissolved oxygen (DO) in the aeration basin.
    b. The mixed liquor suspended solids (MLSS) are too high.
    c. High DO in the secondary clarifier.
    d. All of the above.
60. What is the remedy for filamentous organisms in the secondary clarifier?
    a. Add chlorine to the secondary clarifier.
    b. Increase the dissolved oxygen (DO) to a minimum of 3 mg/L in the aeration basin.
    c. Decrease the ratio of food to microorganisms (F:M) if the mixed liquor suspended solids (MLSS) are too high by wasting sludge.
    d. Check the nutrient conditions, especially the ratio of carbon to nitrogen to phosphorus (C:N:P).

61. What is the significance of a bulk clarifier hydraulic loading rate, such as 2,000 gpd/ft^2?

a. It is a measure of the net fluid upward velocity that controls the speed of mixed liquor suspended solids (MLSS) settling.
b. It controls the depth and stability of the sludge blanket.
c. It controls the overflow rate for the effluent weirs to prevent solids washout over the weirs.
d. It determines the maximum sludge deposition rate.

62. What factors can cause large increases in plant inflow?

a. Rain.
b. Downspouts illegally connected to the storm sewer.
c. Anything that could raise groundwater levels and that increase infiltration and inflow.
d. Firefighting flow drainage if there is a fire in the community.

63. What is pH, and what is the proper range for biological activity?

a. It is a measure of the hydrochloric acid concentration in the water, and the optimal range is between 6.6 and 4.0.
b. It is a measure of the hydrogen ion concentration in water, and the optimal range is between 10 and 12.
c. It is a measure of the acidity or basicity of the water, and the optimal range is between 9 and 6.
d. It is a measure of alkalinity of the water, and the optimal range is 5.5.

64. What is alkalinity, and why is it necessary in the water?

a. Alkalinity is the value created by methanol in the water, and it is needed for plant growth.
b. Alkalinity is the product of bicarbonates and weak acid salts, and it is required for anaerobic degradation and denitrification.
c. Alkalinity is the product of disassociation of calcium hydroxide, and it is required to precipitate metals in the influent.
d. Alkalinity is generated by sulfur dioxide and is used to control the pH.

65. What is an oxidation ditch?

a. It is a primary aeration system.
b. It is a form of extended aeration treatment.
c. It is a special form of treatment for biosolids.
d. It is another name for conventional activated sludge.

66. What happens when the mixed liquor suspended solids (MLSS) in the aeration tank gets too high (above recommended limits)?

a. Aeration becomes more difficult because of the solids inhibiting the oxygen transfer.
b. The MLSS contains more solids, which increases the viscosity.
c. The MLSS concentration can overload the clarifier and cause poor settling.
d. All of the above.

67. What is septage?

    a. Septage is the seeping from a wastewater treatment plant in industry.
    b. Septage is another name for the materials taken from a septic tank or holding tank.
    c. Septage is an aerobic material that contains partially treated fats, oils, greases, and various types of human waste materials.
    d. Septage is the seepage from a landfill disposal site, which is full of nutrients.

68. Why are stabilization ponds also called sludge lagoons?

    a. They are very lightly loaded wastewater treatment impoundments that can provide the equivalent of secondary treatment.
    b. They are an impoundment that encourage natural processes and algae to promote waste treatment.
    c. They are a routine impoundment used in the dairy industry for all wastes and which can be anaerobic or aerobic.
    d. They are a place for waste secondary sludge storage.

69. What is the effect of photosynthesis and respiration on a stabilization pond?

    a. During daylight algae produces O_2 by photosynthesis and raises the pH. At night algae and microorganisms convert the O_2 and give off CO_2, which lowers the pH.
    b. The pH falls as anaerobic processes take over and produce fatty acids.
    c. Sludge does not accumulate but is consumed.
    d. Algae is not an important consideration in the operation of a stabilization pond.

70. Can a stabilization pond treat wastewater just as well as an activated sludge system?

    a. It depends on the amount and strength of the waste being treated.
    b. Stabilization ponds often encourage algal growth that reduces effluent quality.
    c. Stabilization ponds have clarifiers to improve the quality of the effluent.
    d. Stabilization ponds do not treat wastewater as well as an activated sludge system.

71. What is a limitation/are limitations associated with using stabilization ponds?

    a. It is easy to exceed the loading rate and turn the pond anaerobic.
    b. Duckweed and blue-green algae can grow and interfere with the treatment.
    c. Periodically, the sludge must be removed from the ponds and that is very expensive.
    d. All of the above.

72. What are good algae control systems for stabilization ponds?

    a. Fish that eat the algae if the pond is aerobic.
    b. Newer technology using ultrasonic equipment that prevents algae growth.
    c. Pond covers and/or dyeing the water to reduce sunlight.
    d. Phosphate control is a critical issue for controlling algae blooms in stabilization ponds and requires heavy doses of metal salts.

73. What is bulking sludge?

a. Wastewater solids in the clarifier that form large floating clumps.
b. A growth of filamentous organisms that causes poor settling in clarifiers.
c. An anaerobic growth that forms on the surface of the aeration basin.
d. A property of the sludge that makes it clump together and plug the clarifier outlets.

74. What are control techniques for bulking sludge?

a. Chlorination and higher recycle that increase the ratio of food to microorganisms (F:M).
b. Wasting more sludge to lower the F:M ratio.
c. Anaerobic conditions or low dissolved oxygen in the aeration basin, increasing the F:M ratio.
d. Control of grease in the plant.

75. What is a dissolved air flotation (DAF) unit and what function does it serve in a publicly owned treatment plant?

a. It can serve as a substitute for a clarifier.
b. It removes grease and oil from the influent.
c. It aerates the influent and reduces the initial demand on the aerators.
d. It removes particulate material to give the aeration basin a feed, which is all dissolved organics.

76. Which processes use an attached growth rather than a suspended growth system?

a. Trickling filter, rotating biological contactors, and membrane bioreactors.
b. Oxidation ditch and sequencing batch reactor.
c. Contact stabilization, Bardenpho, and University of Capetown system.
d. Orbal waste treatment system and Phoredux system.

77. What do a trickling filter, a biotower, and a rotating biological contactor (RBC) have in common?

a. The processes rely on development of a biofilm attached to a media for stability and support.
b. The residence time in the aeration basin is longer than with an activated sludge plant.
c. The processes rely on surface contact with the water for aeration.
d. All of the above.

78. Compare or contrast surface aeration versus coarse bubble aeration for ponds and lagoons.

a. Both fine and coarse bubble aerators are more efficient at oxygen transfer than surface aerators.
b. Surface aerators can heat the water in ponds and lagoons.
c. Low-speed surface aerators have a lower oxygen transfer efficiency than coarse bubble aerators.
d. Surface aerators can create standing waves that can destroy the embankments on a lagoon.

79. Which of the following is the most true about surface aerators?
   a. They provide good mixing for large areas.
   b. They provide a vertical circular flow pattern but have no oxygen gradient in that flow pattern.
   c. They are less expensive to operate than other types of aeration systems.
   d. The have little or no problems with icing in cold weather.

80. In the provided diagram of an advanced wastewater treatment plant, which of the following is true?

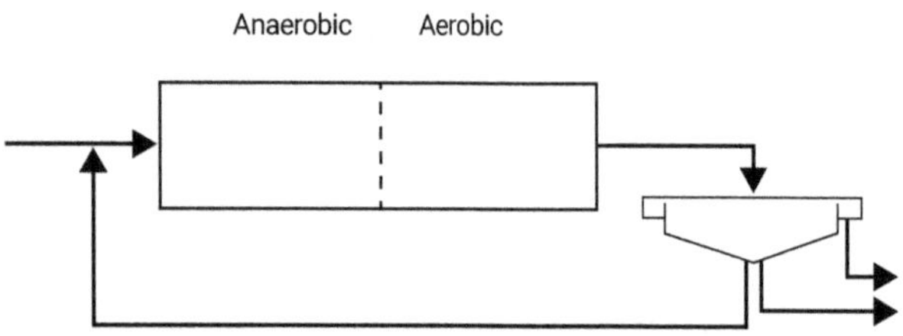

Figure source: *Chapter 16, page 385, 2nd Edition Practical Wastewater Treatment by D.L Russell*

   a. The purpose of the anaerobic section is to reduce and remove both nitrogen and phosphorous.
   b. The purpose of the anaerobic section is to "starve" microbes and make them more efficient at removing contaminants in the aeration section.
   c. There really is no purpose. The anaerobic section is where the return sludge reaches the aeration tank, and because of the load of oxygen demanding microorganisms, the first part of the aeration system will be anaerobic.
   d. The anaerobic section helps to degrade grease and oils.

81. Given the provided figure, what is the purpose of the anaerobic and anoxic tanks?

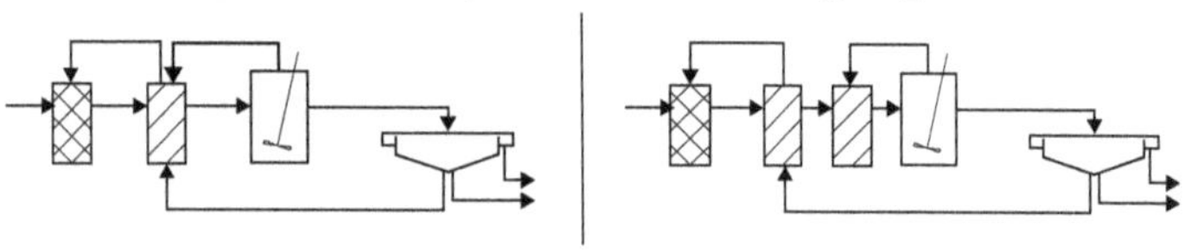

Figure source: *Chapter 16, page 385, Practical Wastewater Treatment, 2nd Edition, by D. L. Russell*

   a. Anaerobic is for phosphorus (P) removal, whereas anoxic is for nitrogen (N) removal.
   b. Anaerobic will remove P and N, whereas anoxic is for additional N removal.
   c. The recycle from the clarifier contains some oxygen, which would poison the anaerobic tank, so the stream is put into the anoxic tank.
   d. The aerobic tank recycle to the anoxic tank is for polishing removal of all N.

82. Given the provided figures, what is the difference in phosphate reduction technologies?

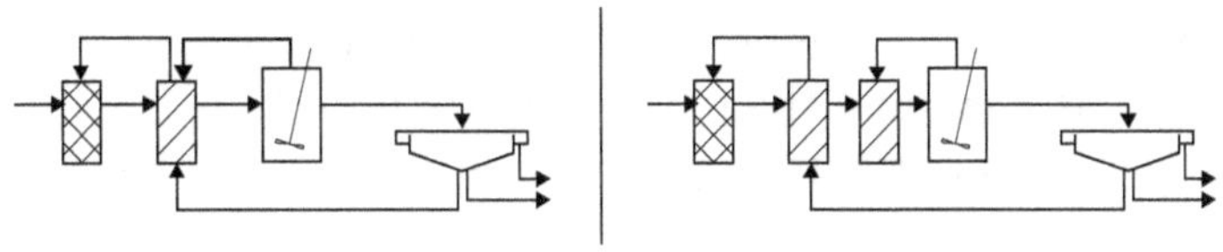

Figure source: *Chapter 16, page 385, Practical Wastewater Treatment, 2nd Edition, by D. L. Russell*

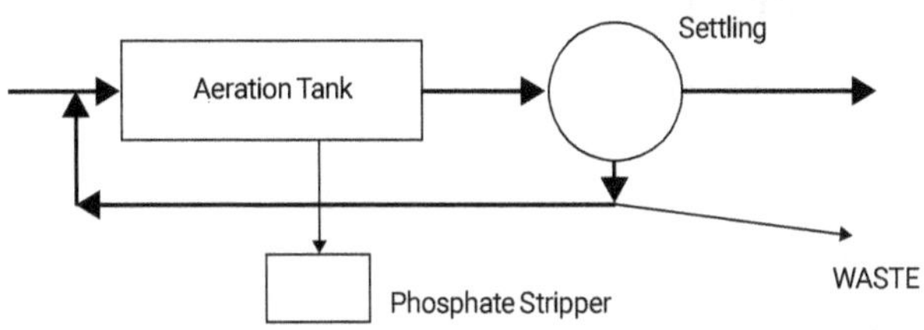

Figure source: *Chapter 16, page 375, Practical Wastewater Treatment 2nd Edition, by David L. Russell*

a. The second figure is a phostrip process where the phosphate is removed by chemical precipitation.
b. The first figure will not remove N and P.
c. The second figure is a biological phosphate-removal system.
d. The first figure will not remove phosphate.

83. What happens when the detention time in the clarifier is too long?

a. The sludge stops settling and starts coming together to form a hard mass.
b. Anaerobic action sets in and starts generating gas bubbles, causing poor performance of the clarifier.
c. The sludge concentration increases to around 10 percent, which can make the sludge hard to pump.
d. The sludge turns acidic, and the acid starts attacking the piping and clarifier internals.

84. Which of the chemicals used for phosphorus precipitation does not generate chemical sludge?

a. Calcium hydroxide
b. Aluminum sulfate
c. Chlorine
d. Ferric chloride

85. What is the approximate biochemical oxygen demand (BOD) and chemical oxygen demand (COD)-to-phosphorus ratios for biological uptake of phosphorus?

a. 20:1 and 45:1
b. 100:1 and 65:2
c. 45:1 and 20:1
d. 35:1 and 40:1

86. What are some of the challenges using metal solutions for precipitation of phosphorus?

a. Corrosion, staining, chemical burns, difficulty in handling, and vapors.
b. US Environmental Protection Agency regulations (Tier II Reporting for Hazardous Chemical usage greater than 500 lb/year).
c. The need for precise dosing and mixing.
d. Potential toxicity from overdosing.

87. What is the significance of the oxygen transfer rate (OTR)?

    a. It is the measure of how quickly oxygen can be transferred into liquid.
    b. It is the rate at which organisms consume the oxygen in the liquid.
    c. It is a direct measure of how much energy is required to maintain aerobic conditions.
    d. It determines the capacity of the aerators.

88. What happens if the oxygen transfer rate (OTR) falls off?

    a. It is an indication of low mixed liquor suspended solids (MLSS).
    b. It is an indication of toxicity.
    c. It means that the aeration basin is not getting as much oxygen.
    d. It means that the aerators are underperforming.

89. What are some of the principal factors affecting the oxygen transfer rate (OTR)?

    a. Bubble size, water temperature, and mixing.
    b. Mixed liquor suspended solids (MLSS) in the aeration basin.
    c. Air temperature of the blower intake.
    d. Amount of recycle in the aeration basin.

90. What is the SOUR (specific oxygen uptake rate)?

    a. It is the same as the oxygen transfer rate (OTR).
    b. It is a measure of specific bacterial uptake of oxygen in the mixed liquor suspended solids (MLSS).
    c. It is a measure of specific bacterial uptake by mixed liquor volatile suspended solids (MLVSS).
    d. It is a measure of the oxygen uptake by the total suspended solids (TSS) in the aeration basin

91. When the specific oxygen uptake rate (SOUR) drops dramatically, what are possible causes?

    a. It is an indication of temperature changes in the mixed liquor suspended solids (MLSS).
    b. It means that the MLSS is lower than desired.
    c. It is a possible indication of low dissolved oxygen (DO) in the aeration tank.
    d. It is a possible indication of toxics in the aeration basin.

92. Which has better performance for biochemical oxygen demand (BOD) and total suspended solids (TSS), and nutrient removals from wastewater, a rotating biological contactor, a trickling filter, or an activated sludge system?

    a. Activated sludge
    b. Trickling filter
    c. Rotating biological contactor (RBC)
    d. Contact stabilization plant

93. What is the similarity between a rotating biological contactor (RBC) and a trickling filter?

    a. Both can be used in a variety of ways to treat wastes and especially industrial wastes.
    b. Both are attached growth reactors that require intermittent dosing or contact with air to provide oxygen to the attached growth.
    c. Both are suspended growth reactors.
    d. Neither require recycle but are efficient for one pass treatment.

94. What is the most common problem (or problems) that causes poor settling of activated sludge in the secondary clarifier?

 a. Growth of filamentous organisms
 b. Low dissolved oxygen (DO) in the clarifier and nitrogen deficiency
 c. Nitrogen and phosphorus deficiency
 d. Hydraulic surge in the plant

95. What is the concern with phosphorus concentrations in publicly owned treatment works (POTW) effluents?

 a. Phosphorus in concentrations as low as 0.008 mg/L can cause algal blooms in reservoirs, and as low as 0.01 mg/L in rivers can cause algal blooms.
 b. Phosphorus can cause algal blooms that supersaturate the receiving water with oxygen during daylight hours but can totally deplete the oxygen at night.
 c. Phosphorus can cause algal blooms that can impart an unpleasant taste or odor into the receiving waters, and certain forms of algae can create cyanotoxins in the waters.
 d. All of the above.

96. If there are concerns about a flowmeter's accuracy, how should it be calibrated?

 a. Use the dilution method by adding a substance upstream at a known rate and then measuring the downstream concentration and comparing the results with the flowmeter.
 b. Check to see that there are no slime buildups on the flowmeter and that there are "critical flow" conditions at the flowmeter.
 c. Use a current meter to measure the velocity upstream and then calculate the flow by velocity area methods.
 d. Calibrate in accordance with the manufacturer's recommendations and check the installation conditions.

97. If one of the sets of distribution arms on a rotating trickling filter plant is worn and needs to be replaced, what special precaution should be taken in servicing the arms?

 a. Wear gloves and boots to prevent infection.
 b. Special personal protective equipment and disposal equipment may be needed to handle the liquid mercury seals if they have to be replaced.
 c. The trickling filter may need to be shut down and decontaminated in the event that there is a hazardous material spill during servicing and replacement.
 d. Special precautions may need to be taken against slips and falls because the media is covered in a zoogleal mass and is quite slippery.

98. Trickling filters and rotating biological contactors have what in common?

 a. They both improve pollutant removal efficiency with recycling from the secondary clarifier.
 b. They both need a primary treatment system.
 c. They both use metallic media to support a biofilm.
 d. They both are suspended growth systems.

99. Which of the following statements are true regarding a wetland being used as a substitute for an activated sludge plant?

    a. The wetland is not designed to handle high quantities of biological matter and solids as in a treatment plant and is not as efficient as a wastewater treatment plant in removal of biochemical oxygen demand (BOD) and total suspended solids (TSS).
    b. The wetland is not a very good nutrient removal system.
    c. The wetland generally requires very large areas for effective treatment, and there is a concern that associated vectors (animals and insects) could spread disease into the community.
    d. The wetland is not regulated by the Corps of Engineers or the state.

100. Can engineered wetlands be used as a polishing system for dechlorinated secondary publicly owned treatment works (POTW) discharges?

    a. No, wetlands should never be used in that manner.
    b. Yes, if there is secondary treatment to remove solids, wetlands are sufficient treatment systems.
    c. Yes, wetlands provide good nutrient removal and can lead to reuse and recycling.
    d. Yes, wetlands can provide good habitat for natural preservation.

101. Why would a state design code limit the weir loading in a secondary clarifier to less than 10,000 gpd?

    a. Because it is related to the surface loading rate and exceeding that would wash solids out of the clarifier by creating an upflow, which counters particle settling.
    b. Exceeding that rate would tend to wash solids near the weirs out of the system.
    c. Exceeding that rate would tend to cause the sludge blanket in the secondary clarifier to destabilize.
    d. To keep from flooding the collection trough.

102. What are some of the challenges and problems with operating a publicly owned treatment works (POTW) with a membrane final filter in lieu of a clarifier?

    a. The membrane system must be cleaned with a citric or other acid solution, and the membranes must be regularly backflushed to prevent solids plugging.
    b. Membranes must be kept moving to prevent slime formations and to prevent plugging.
    c. The bacteria will attack and destroy the membranes.
    d. Hard water can plug the membranes.

103. What are some of the disadvantages of trickling filters?

    a. Low five-day biochemical oxygen demand (BOD5) and chemical oxygen demand (COD) removals as compared to activated sludge
    b. Constant maintenance required on distributor arms and nozzles.
    c. Use of mercury seals on distributor arms.
    d. All of the above.

104. Psychoda flies, telmatoscopus flies, midges, and odors are a particular nuisance for what type of treatment?

   a. Activated sludge drying beds
   b. Trickling filters
   c. Waste treatment ponds
   d. Primary and secondary clarifiers

105. What is the control measure for nuisance *Psychoda alternate, Telmatoscopus albipunctatus,* and midges?

   a. Periodic flooding (submergence) of the system and chlorination
   b. Insect growth regulators
   c. Pesticides
   d. Tank covers

106. What are some of the disadvantages of rotating biological contactors (RBCs)?

   a. Frequent maintenance on shafts, bearings, and seals.
   b. Degradation of the biofilm support medium.
   c. High operating costs with lower removal efficiencies.
   d. All of the above.

107. What are the factors that affect performance of effluent filters such as slow sand filters?

   a. Media size, media distribution, hydraulic loading, and pore size of the openings.
   b. Frequency of backwash when the filter losses reach a critical point.
   c. Formation of a bio layer on the surface of the filter.
   d. Chlorine content at the influent to the filter.

108. What is a microstrainer?

   a. A pretreatment device most commonly used on stormwater.
   b. A rotating screen that is used to remove fine particles from incoming water.
   c. Another name for a parabolic shaped screen with a very fine screen opening.
   d. Another name for a membrane system used for treatment of wastewater in lieu of a clarifier.

109. What are the pretreatment requirements before ultrafiltration or membrane filtration?

   a. Screening or filtration removal of very large abrasive particles and conditioning of the water.
   b. Pre-clarification and removal of mixed liquor suspended solids (MLSS).
   c. No special pretreatment required.
   d. Chemical addition to coagulate particles.

110. What are the three types of configurations for a membrane filtration system?

   a. Plate and frame, spiral wound, and spaghetti strand.
   b. Packed, pleated, and fiber reinforced.
   c. Tubular, string-like, and microballs.
   d. Compressible porous balls (fuzzy filter).

111. How can one tell if a plant is biologically overloaded?

a. By the color of the mixed liquor suspended solids (MLSS) in the aeration basin.
b. By extremely low (less than 1 mg/L) dissolved oxygen (DO) even when aerators are running at their maximum capacity.
c. By odors coming off the aeration basin.
d. By the presence of very viscous looking sludge and high levels of sludge in the final clarifier at normal operational levels.

112. What is the difference between a plug flow and a complete mix treatment system?

a. A complete mix treatment system introduces waste at various points along the system treatment train, whereas a plug flow system treats everything in sequential order.
b. A plug flow system introduces new material halfway through the aeration basin.
c. It makes no difference.
d. A complete mix system has a much higher recycle rate

113. What are the cycles in a sequencing batch reactor?

a. Fill, aerate, rest, and decant.
b. Fill, react, settle, decant, and rest.
c. Fill, react, decant, sludge draw-off, and refill.
d. Aerate while filling, react, rest, decant, and refill.

114. What is pH?

a. pH is a measure of chemical activity.
b. pH is a measure of alkalinity of the wastewater.
c. pH is used to measure acid strength.
d. pH is a measure of hydrogen ion concentration on a logarithmic scale.

115. What is the pH of an alkaline solution?

a. Greater than 7
b. Between 5 and 9
c. Below 7
d. At 7.0

116. What is the difference between chemical oxygen demand (COD) measured by the dichromate method and the COD measured by the potassium permanganate ($KMnO_4$) method?

a. The $KMnO_4$ method gives inconsistent results and is no longer used.
b. The dichromate method gives much higher results than the $KMnO_4$ method.
c. The dichromate method requires careful disposal of the test materials because of the presence of mercury and hexavalent chromium.
d. The accuracy of the dichromate method is still in dispute.

117. What is the pH of an alkaline solution?

a. >7
b. 5–9
c. <7
d. 7.0

118. How does one remove excess ammonia from influent wastewater?

a. Raise the pH to above 8.3 and aerate.
b. Add caustic to raise the pH to between 10 and 12 and aerate.
c. Add excess chlorine to reduce ammonia to nitrogen gas (N_2).
d. Oxidize ammonia to nitrate in the aeration tank if possible by adding more mixed liquor suspended solids (MLSS).

119. What are the principal forms of nitrogen in the water environment?

a. N_2O_5, NH_3, NO_2
b. NH_3, NO_3, N_2O_5
c. NH_3, NO_3, N_2O_5, NO_2
d. N_2, NH_3, NO_3, HNO_3, NO_2

120. Gas chlorine and sulfur dioxide (use for dichlorination) should

a. be stored in a separate building, away from any fuels or combustible materials.
b. have easily accessible SCBA (self-contained breathing apparatus) gear and warning signs outside of the building where they are stored.
c. be listed on an Emergency Planning and Community Right-to-Know Act (EPCRA) Tier II disclosure form if present in reportable quantities.
d. be stored in separate rooms with adequate ventilation.

121. Which one of the following chemicals will not remove phosphorus?

a. Calcium hydroxide
b. Aluminum sulfate
c. Sulfur dioxide
d. Ferric chloride

122. What causes the majority of smells associated with anaerobic digestion/anaerobic treatment?

a. Sulfur dioxide
b. Dihydrogen monoxide
c. Hydrogen sulfate
d. Hydrogen sulfide

123. What is one of the best ways of suppressing odors at a treatment plant?

a. Keep the aeration system operating well at 4 mg/L dissolved oxygen (DO).
b. Cover the aeration tanks and vent to a tall stack.
c. Collect the air and process through an incinerator
d. Collect the air and use an alkaline scrubber for the worst streams.

124. What is the sludge volume index (SVI) indication of a good settling sludge?

a. The SVI of greater than 1,000 is a measure of a good settling sludge.
b. The SVI of greater than 500 is a measure of a good settling sludge.
c. The SVI of less than 150 is a measure of a good settling sludge.
d. The SVI of greater than 150 is a measure of a very compact settling sludge that is probably too old.

125. Which test is used to measure chemical oxygen demand (COD)?

    a. A colorimetric test involving reagents
    b. An incubation test over 5 days
    c. A caustic digestion followed by colorimetric testing
    d. A digestion in potassium dichromate followed by colorimetric measurement

126. How is the soluble five-day biochemical oxygen demand (BOD5) test different from the regular BOD5 test?

    a. The non-soluble BOD is first precipitated and then the soluble BOD is measured.
    b. The soluble BOD is filtered out by a 0.45-micron filter.
    c. There is no difference between soluble and insoluble BOD.
    d. For the soluble BOD5 test, particulate material is filtered out by a 0.45-micron filter, and the test is run as a normal BOD5 test.

127. Why is it important to know both the soluble and insoluble phosphorus content in sludge?

    a. The sludge is recycled to the head end of the plant for phosphorus control.
    b. The sludge filtrate is recycled to the aeration basin of the plant.
    c. The insoluble phosphorus stays with the sludge and can be used for land application.
    d. The soluble phosphorus can be used as a liquid fertilizer.

128. What is the process significance of measuring oxygen-reduction potential (ORP) rather than dissolved oxygen (DO) in an anoxic/anaerobic environment?

    a. Either can be used to determine the difference between anoxic and anaerobic.
    b. The DO probes cannot distinguish between anoxic and anaerobic environments where there is very low to no DO.
    c. In an anoxic environment, the ORP is slightly negative (–50 mv), and in an anaerobic environment the ORP is much more negative (even less than –100 mv)
    d. The anoxic environment and the anerobic environment both have zero DO.

129. What types of organisms are visible under a microscope in mature mixed liquor suspended solids (MLSS)?

    a. Free swimming ciliates, flagellates, and amoeboids
    b. Stalked ciliates, free swimming ciliates, flagellates
    c. Nematodes, rotifers, stalked ciliates, free swimming ciliates, amoeboids
    d. Water nymphs, celluliters, bugs, and ciliates

130. What is the difference between analytical accuracy and analytical precision?

    a. No difference
    b. Accuracy is the true value of the substance, whereas precision is the repeatability of the answer.
    c. Precision is the process of getting the same result several times.
    d. Precision is not necessarily related to accuracy.

131. What are the chemical reference standards for total suspended solids (TSS), mixed liquor suspended solids (MLSS), and mixed liquor volatile suspended solids (MLVSS)?

    a. 100 mg/L, 1,000 mg/L, and 800 mg/L
    b. There is no standard for any solids measurement.
    c. 25 mg/L, 500 mg/L, and 400 mg/L
    d. 5 mg/L, 50 mg/L, and 100 mg/L

132. What is the difference between the measurement technique for mixed liquor suspended solids (MLSS) and mixed liquor volatile suspended solids (MLVSS)?

a. MLSS is dried at 83°C and MLVSS is dried at 426°C.
b. MLSS is dried at 105°C and MLVSS is ignited at 550°C.
c. MLSS is dried at 150°C and MLVSS is dried at 550°C.
d. MLVSS is dried at 110°C, then ignited at 1,000°C.

133. After a 0.42-micron membrane secondary filter system, what is the total suspended solids (TSS) in the liquid after filtration?

a. Between 5 mg/L and 10 mg/L
b. Less than 5 mg/L but greater than 3 mg/L
c. Less than 1 mg/L
d. Between 1 mg/L and 3 mg/L

134. What are the recommended temperature ranges for anaerobic digestion?

a. Psychrophilic: 33–68°F
b. Mesophilic and thermophyllic: 95–145°F
c. Superthermophyllic: 160–200°F
d. Mediophyllic: 70–90°F

135. What effect does increased temperature have on anaerobic digestion?

a. Increased solids digestion
b. Increased gas production
c. Shorter residence time
d. All of the above

136. What is the difference in biosolids classification produced by mesophilic and thermophilic digestion?

a. None
b. Mesophilic produces class B biosolids.
c. Thermophilic produces class A biosolids.
d. Thermophilic produce class C biosolids.

137. What are the advantages of mesophilic digestion?

a. High sludge feed concentrations and acceptable gas production.
b. High energy input and it generates class C biosolids.
c. It has a shorter holding time (60–95 days) and smaller tankage than psychrophylic digestion.
d. Mesophilic gas production is about the same as psychrophilic gas production.

138. What are the advantages of thermophilic digestion?

a. Short residence time, class A biosolids, and a high rate of gas production.
b. Larger sludge processing rate and larger tank volumes, but poor-quality gas production.
c. Much higher energy requirement (120°F minimum) and a much wetter digested sludge.
d. No advantage as compared to mesophilic digestion.

139. What is the purpose of a belt filter press?

   a. Increases solids content in sludge.
   b. Removes solids before anaerobic digestion.
   c. Dewaters the waste activated sludge before digestion.
   d. Dewaters sludge before sending it to the drying beds.

140. New technology has improved the treatment of anaerobic digestion processes. These improvements include

   a. Development of maximum recycle anaerobic digesters.
   b. Development of two stage (multi-compartment) anaerobic digesters.
   c. Development of high solids anaerobic digesters.
   d. Development of an anaerobic-aerobic digester system

141. What are uses for digester gas?

   a. A processed digester gas is generally suitable for combustion engines.
   b. Digester gas contains a lot of water and carbon dioxide (CO_2), ammonia (NH_3), and hydrogen sulfide (H_2S), and has a low fuel value if used directly.
   c. Unprocessed digester gas is not harmful or corrosive to combustion equipment.
   d. Unprocessed digester gas has a very high CO_2 content and contributes to greenhouse gas load more than methane.

142. What is the difference between a filter press and a belt filter?

   a. The filter press dewaters sludge by compressing it between plates, whereas the belt filter compresses and shears the sludge as it wraps around rollers.
   b. The belt filter is difficult to operate, whereas the filter press is not.
   c. The filter press can produce a dryer sludge cake than the belt filter.
   d. The belt filter is a preconcentration step before the filter press.

143. What can an operator do to increase the solids coming off the belt filter?

   a. Dilute the sludge to sterilize the sludge.
   b. Add polymers that cause the sludge feed solids to increase.
   c. Adjust the rollers on the belt filter to decrease their tension.
   d. The operator is relatively unable to increase the solids concentration coming off the belt filter without redesign and remanufacture of the system.

144. What is the most environmentally friendly use for sludge?

   a. Land spreading to incorporate it into the soils.
   b. Convert it into compost for a soil amendment.
   c. Incinerate it and destroy it.
   d. Landfill it.

145. What contaminants might be in sludge that could prevent land application?

   a. Chromium, copper, zinc, and iron
   b. Oil and grease
   c. Arsenic, mercury, molybdenum, cadmium, and selenium
   d. Iron, lead, calcium, and phosphate

146. What is the purpose of sludge filtration?

a. Reducing the volume, storage, and transportation costs.
b. Eliminating free liquids before landfill disposal.
c. Reducing fuel requirements if residuals are to be incinerated or dried.
d. Making the sludge easier to spread (less viscous).

147. Based on temperature, what is the normal hydraulic detention time (HRT) required for anaerobic digestion?

a. At 35–38°C, it is between 15 and 40 days, depending on average temperature.
b. At 55–60°C, it is can be greater than 15 days but less than 40 days.
c. At 10°C and below, the HRT is 30 days.
d. At 80°C, and above, the HRT is at least 90 days.

148. What are the operating requirements for aerobic digestion?

a. Minimum dissolved oxygen (DO) requirements are 0.4–0.8 mg/L.
b. Volatile acids/alkalinity ratio of about 10.
c. Temperature of at least 20°C.
d. Mixed liquor suspended solids (MLSS) not exceeding 3,500 mg/L.

149. What do siloxanes do in anaerobic digestion?

a. Siloxanes are compounds found in sealants, adhesives, coatings, plastics, cosmetics, medical devices, food contact materials, and many other industrial applications and do not interfere with anaerobic digestion.
b. Siloxanes volatilize in gases and interfere with combustion of the generated gas by turning into glass.
c. Siloxanes interfere with anaerobic digestion.
d. Siloxanes decrease the energy value of the waste gas.

150. What is struvite and what does it do?

a. It is a natural compound that is formed in anaerobic digestion, which can coat the mixers on the inside of the digester.
b. It is a substance commonly found in kidney stones as a result of bacterial infection.
c. It is an ammonium, magnesium, calcium compound that can create deposits in pipes and valves, as well as in the digester itself.
d. It is a byproduct of anaerobic digestion that can form stones and crystals under the right circumstances.

151. What factor(s) is/are important in biosolids production?

a. Acceptable metal concentrations and pathogen reduction
b. Good nutrient values
c. Reduction of attractiveness to vectors
d. Suitable form for easy handling and spreading

152. Method(s) of producing a class A biosolid include which of the following?

a. Composting, thermal treatment, or lime stabilization
b. Lime stabilization
c. Thermal stabilization
d. Aerobic digestion of sludge

153. How does liquid sludge disposal compare to dewatering?

a. Liquid sludge disposal is easy to apply but is hauled in tankers, whereas dewatered sludge has higher capital and operating cost and lower hauling costs (trucks versus tankers).
b. Sludge dewatering is more environmentally friendly.
c. Liquid sludge disposal can be conducted year-round.
d. Dewatered sludge cannot be incorporated into land.

154. Land application of sludge in liquid or solid form should be

a. analyzed and applied at agronomic rates so that plants will take up excess nutrients.
b. injected deeper into the soils (3 ft or more).
c. carefully analyzed for metal concentrations, which might be detrimental to crops.
d. plowed into the soils at depths of 6 in. or more.

155. What are the advantages of a filter press (recessed plate filter) over a belt filter?

a. Filter presses produce a drier sludge cake.
b. Replacing the filter elements (e.g., cloth filters) is easier with a filter press than with a belt filter.
c. The cycle time is less with a filter press for a selected cake solids concentration compared with a belt filter.
d. A filter press requires less cleaning to remove sludge than a belt filter.

156. In an anaerobic digestion process, what is likely to occur?

a. The organic matter is hydrolyzed and made soluble and then fermented into acetic acids, hydrogen, and carbon dioxide (CO_2), and then into methane.
b. The organic matter is converted directly into methane and CO_2.
c. The organic matter is degraded by aerobic and anaerobic action to CO_2.
d. The organic matter reacts with the existing organisms and breaks down aerobically.

157. The end products of anaerobic digestion are

a. methane and CO_2.
b. inert biosolids that cannot be degraded further.
c. a wet sludge, gas ($CH_4 + CO_2$), acids, and liquid.
d. a volatile and stable fuel.

158. The principal bacteria (organisms) in anaerobic digestion are composed of ______________.

a. methanogens
b. acid formers
c. both acid formers and methanogens
d. facultative bacteria but not strict anaerobes

159. What is the optimal pH for a digester to operate?

a. 6–9
b. 6.5–7.5
c. 7–10
d. A digester will operate at almost any pH.

160. What are the steps to convert biochemical oxygen demand (BOD) to carbon dioxide (CO_2) and methane?

a. Volatilization, conversion to organic acids, and conversion of acids to methane and CO_2.
b. Hydrolysis, conversion to organic acids, and conversion of acids to methane and CO_2.
c. Bacterial conversion of volatile organics into organic acids (e.g., acetic acid) and propionic acid, and then methane formers convert the acids into methane and CO_2.
d. Enzyme conversion of volatile organics directly to methane and CO_2.

161. If a digester gets "stuck" in a "sour" cycle under normal operation, what is the probable cause?

a. Industrial wastes in the sludge.
b. Too much new sludge fed to the digester beyond what the prokaryotic organisms can use, causing the pH to drop.
c. Aerobic materials have been fed to the anaerobic digester.
d. The mixing in the digester has failed.

162. What are volatile fatty acids (VFAs) and why are they important?

a. VFAs are biopolymers that promote coagulation.
b. VFAs are the food that methanogens (PAOs) use to produce methane and carbon dioxide (CO_2) and they accumulate phosphorus.
c. VFAs are produced by PAOs to precipitate the excess phosphorous in the mixed liquor suspended solids (MLSS).
d. VFAs are byproducts of the uptake of phosphorus by PAOs.

163. What is the effect of too much volatile fatty acid production?

a. Bacterial inactivation caused by an increase in pH.
b. Reduction in metabolic activity as a result of toxin production.
c. Methanogenic inactivation caused by low pH.
d. Accumulation of flammable volatile gases in the digester.

164. What should be done to address a stuck or "sour" digester?

a. Add calcium carbonate and lime carefully to raise the pH.
b. Add calcium carbonate (only) to raise the pH.
c. Add sodium hydroxide to raise the pH.
d. Add quicklime to raise the pH.

165. What should be done with the liquid that is left over after dewatering anaerobic digestion sludge?

a. Put it back in the head end of the plant for further processing.
b. Spread it on sand beds for drying.
c. Put it into a holding pond.
d. Reprocess it by putting it back in the anaerobic digester.

166. How can lime be used to stabilize sludge?

   a. The addition of free lime can be used to raise the pH of the sludge to 8.
   b. The pH of the limed sludge must be raised to 10–12.
   c. Depending on the process, the sludge must be held at a high pH for two weeks.
   d. Quick lime should not be mixed with sludge because it will catch fire.

167. What is composting?

   a. A method of disposing of sludges through anaerobic degradation.
   b. A method of creating a soil amendment through aerobic degradation.
   c. A waste disposal technique for all solids.
   d. A mix of sludge, leave or grass, and wood chips used in gardening.

168. What are the safety hazards associated with manholes?

   a. A manhole is a confined space as defined by the Occupational Safety and Health Administration (OSHA).
   b. The employees can trip or injure themselves in the confined space.
   c. It is difficult to access manholes safely with ladders.
   d. Manholes are difficult to ingress and egress safely because of the weight of the manhole covers and the likelihood that they could fall and injure someone in the manhole.

169. Why is it important to wear a life jacket when working around aeration basins?

   a. The turbulence and aeration can pull a person under the water.
   b. The density of aerated wastewater is less than the density of the human body, and it is impossible to swim out.
   c. If an individual falls into an aeration basin but is wearing a life preserver, they may drown even if they can swim.
   d. The aeration currents generated by rising bubbles can pull a person under water even if they are wearing a life preserver.

170. Why should gloves, lab aprons, and goggles be mandatory when handling lab chemicals and wastewater?

   a. To prevent contamination of the liquid samples.
   b. For worker protection from infection, spills, and splashes.
   c. To prevent chemicals from eating through work clothes.
   d. To enable employees to handle corrosive chemicals safely.

171. Where should reagents and flammable materials be stored?

   a. In a leak proof cabinet and flammables cabinet.
   b. In separate spill proof and/or flammables cabinets.
   c. It really doesn't make any difference if chemicals are stored carefully.
   d. Storage is not the issue, spills of the chemicals are of concern.

172. Municipal employees are covered by and must work in compliance with the Occupational Safety and Health Administration (OSHA) rules.

   a. No.
   b. Yes, but OSHA authority is limited.
   c. Yes, but only those parts of the OSHA regulations that apply.
   d. Municipal workers are exempt, but outside contractors are fully covered by the regulations.

173. What should be done at a facility in the event of a spill from a heating oil tank or diesel fuel tank that holds more than 660 gal and is aboveground?

   a. The tank should be on a concrete pad and spilled oil drained to the aeration basin of the treatment works.
   b. In the event of a discharge of oil, the discharge should be reported to the state Environmental Protection Agency (EPA) and USEPA within 24 h.
   c. The spill should be cleaned up as well as possible but doesn't need to be reported to anyone.
   d. The US Coast Guard should be called to clean up the spill.

174. What should be done in the event that a facility is notified that a gasoline truck has overturned and a large slug of gasoline drained to the sanitary sewer?

   a. The fire department should be notified to be on standby.
   b. Plant employees should be warned, and if necessary, the plant should be evacuated until the spill passes.
   c. Employees should be alerted to vapor/fire/explosion hazards and not to smoke because the fuel will evaporate in the aeration basin.
   d. The gasoline flow should be bypassed to avoid the treatment plant.

175. What should be done before a direct drive pump is disassembled and the motor is serviced, considering that the direct drive pump is a lead pump in a series of pumps that has been turned off, and the impeller is still spinning as a result of the force of fluids through the pump.

   a. Close the inlet and outlet valves to stop the flow through the pump volute.
   b. Disconnect the pump coupling from the motor.
   c. Disengage the breakers or fuses in the box and throw the main switch to the off position.
   d. Deenergize the motor by disconnecting the electrical leads.

176. If an occasional bypass is identified on a wastewater treatment plant, does it need to be reported to the regulatory agency and, if so, how quickly?

   a. It must be reported within 24 h verbally to the state or federal government, and a written report must be filed within five days. The report must include an analysis of the causes and measures taken to prevent recurrence.
   b. It must be included in the plant's monthly monitoring report with analysis of the causes and measures taken to prevent a recurrence.
   c. It must be reported within 24 h with an explanation of the cause and the proposed remedial activity.
   d. It's part of routine operations and does not need to be reported.

177. In performing compliance sampling on someone discharging to a system, what elements should be included in the written sampling program?

   a. Sampling protocols, a discussion of analytical methods, the method for sample preservation, the number and kind of samples collected, and a list of containers and their contents.
   b. Everything listed in (a) plus the sample marking and labeling protocols.
   c. Everything listed in (b) plus the required sample volume and the protocols for split sampling with the industry.
   d. The samples should be labeled for the laboratory with time, date, and analyses required.

178. What is the purpose of disinfection?

a. To kill all the pathogens in wastewater.
b. To kill any remaining bacteria and spore-forming organisms.
c. To purify the water and make it potable to use for drinking.
d. To treat the water to reduce the likelihood of algae growth in receiving waters.

179. What is the recommended optimal detention time for a chlorine contact tank at peak flows?

a. 45 min
b. 30 min
c. 15 min
d. 5 min

180. What are the most commonly used disinfecting agents used for wastewater treatment?

a. Cl_2, NH_3
b. $HClO^{-}$, $OCl^{=}$
c. Ozone
d. NaClO

181. What is breakpoint chlorination?

a. Application of chlorine until its odor can be detected.
b. Application of chlorine to a level that breaks up all the bacteria.
c. Application of chlorine until it oxidizes all amines and is present as free chlorine.
d. Application of chlorine until it disinfects all the suspended solids.

182. Why should effluent be dechlorinated before discharge?

a. Aesthetic reasons, principally odor.
b. Aquatic toxicity of chlorine is 37 mg/L for many species.
c. Chlorine is a persistent chemical and can be tasted in drinking water streams.
d. Chlorine will off-gas in the environment and contribute to ozone depletion.

183. What chemical(s) are commonly used for dechlorination?

a. Ammonia and ammonium hydroxide
b. Sulfur dioxide and sodium bisulfite
c. Calcium hydroxide
d. Soda ash

184. What are some of the challenges in working with ultraviolet (UV) disinfection?

a. Disinfection depends on transmittance of the fluid, intensity of radiation, and time of exposure. Even if properly treated, organisms can still regrow.
b. Lamp fouling and bacterial growth will not an affect disinfection performance.
c. Detention times are very long.
d. Lamps are expensive and have an extremely short life.

185. What are the challenges of working with chlorine?

a. Chlorine is a corrosive and toxic gas. If inhaled, it can be fatal.
b. Chlorine is a toxic gas that is difficult to transport and can represent a danger to the community when it is stored in bulk or accidentally released or mishandled.
c. Chlorine is not as effective a disinfectant as ozone.
d. Chlorine reacts with ammonia to form chloramines, which are less effective disinfectants.

186. What are the challenges of working with sodium hypochlorite?

a. It is a strong solid oxidizer and can cause burns.
b. It requires safe handling and special personal protective equipment (PPE).
c. It can start to decompose above 40°C (104°F), releasing chlorine.
d. It can safely be stored in metal containers.

187. What are the challenges associated with ozone as a disinfectant?

a. It is a hazardous gas that cannot be stored, must be generated on site, and is highly corrosive as an oxidizer.
b. Casual exposure is no problem.
c. It is chemically oxidative but can be stored in steel.
d. It condenses to droplets at temperatures below 20°F.

188. The effectiveness of a disinfectant is determined by ________________.

a. concentration
b. contact time
c. log reduction in viable organisms
d. concentration, contact time, and dose

189. Which are harder to disinfect in wastewater?

a. Bacterial spores
b. Mold spores
c. Viruses and bacteriophages
d. *Escherichia coli*

190. Which is the more powerful disinfectant?

a. Chlorine
b. Chlorine dioxide
c. Ozone
d. Ultraviolet light

191. What are the challenges with the use of ozone as compared with chlorine?

a. Ozone cannot be stored but must be generated at the time of use whereas chlorine can be stored.
b. Ozone is significantly less expensive than chlorine.
c. Ozone does not leave a residual in the water, and it is easy to re-contaminate the water supply.
d. Ozone leaves a distinctive taste in the water.

192. What is peracetic acid (PAA)?

a. A very effective disinfecting compound made by reacting acetic acid and hydrogen peroxide.
b. An expensive disinfectant used in the food industry.
c. A less powerful disinfectant than chlorine dioxide.
d. A disinfectant that is being experimentally used as a replacement for chlorine.

193. What, if any, are the benefits of using peracetic acid (PAA) as a disinfectant?

 a. It doesn't generate disinfection byproducts.
 b. It is as powerful as sodium hypochlorite (NaOCl).
 c. It is not as dangerous to handle or as corrosive as NaOCl.
 d. All of the above.

194. What are chloramines?

 a. Disinfection byproducts formed when chlorine reacts with organic compounds.
 b. A byproduct of the reaction between chlorine and ammonia.
 c. A powerful disinfectant from the reaction of ammonia and chlorine.
 d. A poor disinfectant with a long life and which is hard to get rid of.

195. What is breakpoint chlorination?

 a. The application of excess chlorine to reduce all ammonia to nitrogen gas (N_2).
 b. The application of chlorine to break down the residual organics.
 c. A way to eliminate disinfection byproducts.
 d. All of the above.

196. What is a weir?

 a. A flow measurement device (a flowmeter).
 b. A flow regulation device.
 c. A thin plate used to increase aeration.
 d. A plate used to control or redirect the direction of flow in a channel.

197. What are the most common types of open channel flow measurement devices?

 a. Parshall flume and variations
 b. Mag meter
 c. Gauging station
 d. V-notch weir

198. What is a Parshall flume?

 a. If properly installed, it is an accurate flow measurement device for closed conduits.
 b. If properly installed, it is an accurate measuring device for open channels.
 c. It is a flowmeter device that needs calibration.
 d. It is a flowmeter that also increases aeration.

199. What are the most common types of open channel flowmeters that are inserted in sewer lines?

 a. Palmer-Bowlus flumes
 b. Parshall flumes
 c. Weirs
 d. Sonic flow measurement devices

200. What are the principal concerns in installing a flowmeter?

 a. Uniform flow upstream of the flowmeter, where the uniform flow is quiescent.
 b. Changes in direction immediately above the flowmeter.
 c. Free surface (as opposed to pipe flow) and free downstream conditions.
 d. Approach water velocity.

201. How accurate are the various types of electronic flowmeters?

   a. 10–15 percent
   b. 5–10 percent
   c. 2–5 percent
   d. Impossible to determine without calibration

202. What are the most common types of flowmeters used in closed pipes, such as with pumping stations?

   a. Venturi meter
   b. Orifice meter
   c. Magnetic flowmeter (magmeter)
   d. Ultrasonic flowmeter

203. What are the potential problems associated with full pipe flowmeters, such as magnetic flowmeters and ultrasonic flowmeters?

   a. Highly variable velocity profiles
   b. Pipe partially full
   c. Uneven or nonuniform flow distribution
   d. High solids in the liquid

204. What is an airlift?

   a. A way of resupply involving airplanes.
   b. A way of collecting sludge in small systems.
   c. A way of using compressed air in the bottom of a pipe as a pump to move water.
   d. A mechanical device for removing heavy aerator piping from the tank.

205. What is an example of a positive displacement pump?

   a. A plunger style pump that is often used to pump sludge.
   b. A submersible pump that removes wastewater from a wet well in a positive manner.
   c. A centrifugal pump with vanes on the casing to create a positive displacement.
   d. A vortex pump is an example of a positive displacement pump.

206. What is a lobed blower?

   a. A positive displacement blower that uses figure-8-shaped rotating impellers to provide positive displacement and move air.
   b. A regenerative blower named after William Lobe.
   c. A blower used to create a high vacuum or high pressure.
   d. A blower that is designed to deliver a constant volume at pressure.

207. What is an Archimedes' screw (also called an inclined screw)?

   a. An old-style, tamper-proof fastening device used on manholes and other secure locations.
   b. A screw used to clean out plugged pipes.
   c. An ancient way of moving water from one level to a higher level.
   d. A low-pressure, continuous cavity pump.

208. What is the difference in performance between a lobed or positive displacement blower, a centrifugal blower, and an axial blower?

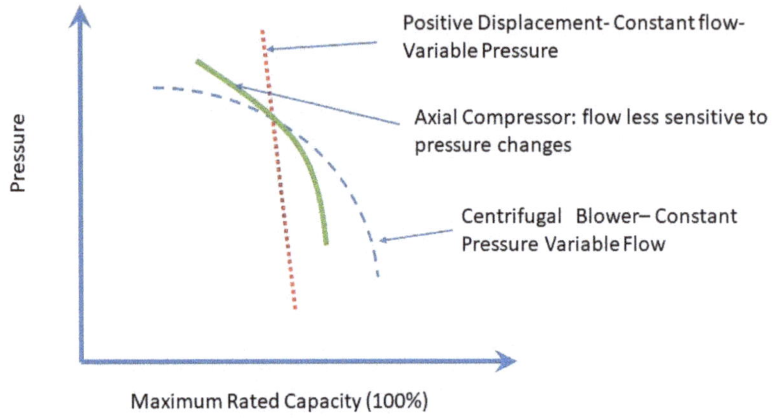

Figure source: *Created by author for this publication*

a. Centrifugal blowers' and axial compressors' capacities decrease markedly with increasing pressure, whereas a positive displacement blower's capacity is relatively independent of pressure.
b. Positive displacement blowers have a highly variable capacity at a given speed and are affected by changes in pressure, and they are often unsuited to wastewater treatment applications compared with centrifugal and axial blowers.
c. Axial blowers have a relatively constant capacity at a fixed speed but the capacity decreases with increasing pressure, although to a lesser extent than with centrifugal blowers.
d. There is little difference between positive displacement, axial, and centrifugal blowers.

209. For a centrifugal blower, what is the shutoff pressure?

a. The pressure at which the blower stops running
b. The pressure that is the maximum the blower can produce
c. The pressure against which there is no more flow through the blower
d. The pressure where the inlet and outlet pressures are identical

210. For a centrifugal pump (incompressible liquid), what it the greatest danger of operating the pump with the inlet and discharge valves closed?

a. The pump will stall.
b. The pump motor will burn out.
c. The pump can overheat and turn the water into steam, damaging to the pump.
d. If left running long enough, the pump will overheat and explode.

211. What is the problem with operating a centrifugal pump with little or no back pressure or resistance?

a. The pump flow will greatly increase beyond design capacity and could cavitate, causing internal damage.
b. The system pressure will drop.
c. The pump motor could burn out because it is overloaded.
d. The pump seals will be damaged by the low pressure.

212. What is the purpose of a motor coupling?

a. It transmits power from the motor to the pump.
b. It can compensate for minor shaft misalignments.
c. It enables separate maintenance of the pump and the motor.
d. All of the above.

213. What is a packing gland on a pump?

a. A split ring that enables one to tighten the packing on the pump to prevent leaks around the shaft.
b. A seal between the pump and the motor.
c. A split ring used to seal the inlet and outlet of a pump from leaks.
d. A place to store excess oil for lubrication.

214. Why is seal water used in a sewage pump?

a. It is used to clean off the mechanical seals.
b. It is used to test the seal between the pump and the piping for leaks.
c. It is used to flush the pump seals from the outside, keeping them clear of grit and dirt, which could shorten their life.
d. In modern pumps, seal water is not necessary.

215. A submersible pump should never

a. be placed on the bottom of a wet well.
b. be allowed to run dry.
c. be hard piped into position.
d. be laid on its side to operate.

216. When installing a centrifugal pump, one should

a. pay attention to the direction of rotation on the pump casing, the electrical connections, and the voltages.
b. wire the motor correctly.
c. ensure that the coupling is connected to the shaft.
d. inspect the seal water connection.

217. What is the effect of running a pump at a higher RPM?

a. Greater flow, lower power demand.
b. Power demand increases with the cube of the RPM ratios.
c. Head increases with the cube of the RPM ratios.
d. Volume increases with the square of the RPM ratios.

218. What is the problem with starting or stopping a pump quickly, or opening or closing a valve quickly?

a. Stopping a pump quickly can cause the motor to bind.
b. Starting a pump quickly can cause the circuit breakers to trip.
c. Closing a valve or stopping a pump can cause a transient pressure spike upstream and downstream of the system.
d. Starting a pump or opening a valve quickly can cause a downstream transient pressure spike.

219. What is total dynamic head on a pump?

   a. The sum of the friction losses on the piping.
   b. The sum of the friction losses plus the suction lift on the piping.
   c. The sum of the friction losses, the suction lift, the entrance and discharge losses on the piping.
   d. The total system back pressure.

220. Besides friction, what determines the maximum height water can be sucked from a reservoir to a pump? Assume a perfect vacuum.

   a. Pump efficiency
   b. Pump elevation
   c. Size of piping
   d. Atmospheric pressure and elevation

221. What is maximum pump suction lift (often called net positive suction head or NPSH)?

   a. The difference between atmospheric pressure and the lift and dynamic head losses on the suction side of a pump.
   b. The maximum theoretical distance a pump can lift water.
   c. The vapor pressure of water under suction.
   d. A value that depends on the atmospheric pressure, the liquid temperature, the vapor pressure, and the suction lift of the pump.

222. What is the shutoff head of a pump?

   a. The maximum pressure a centrifugal pump can produce at zero flow.
   b. The pressure at which one should shut off the pump to avoid damage.
   c. The pressure at which the pump should be shut down to avoid damaging the motor.
   d. The maximum system pressure.

223. Starting or stopping a pump quickly should be avoided to prevent

   a. burning out the pump motor by starting it under load.
   b. creating pressure surges, which can damage other pumps in the system.
   c. cavitation in the pump impeller.
   d. creating a huge draw on the electrical system and possibly tripping a circuit breaker.

224. A cutter or chopper pump is used when

   a. there are rags and clogs in the flow.
   b. there are large solids in the flow.
   c. the application is for a submersible pump.
   d. there is a need for a pump for very rough service.

225. A vortex pump

   a. has a recessed impeller design.
   b. is more efficient than a comparable centrifugal pump.
   c. cannot pump fluids containing larger solids than a centrifugal pump.
   d. is a good application in which there is grease in the wastewater.

226. A diaphragm pump is most often used when pumping what?

a. Viscous liquids and chemical slurries, corrosive chemicals, sludge, and mud.
b. Return solids from the clarifier underflow.
c. The recycle from the aeration basin to the head of the plant.
d. Wastewaters containing large and occasionally stringy solids.

227. Identify what is depicted in the image.

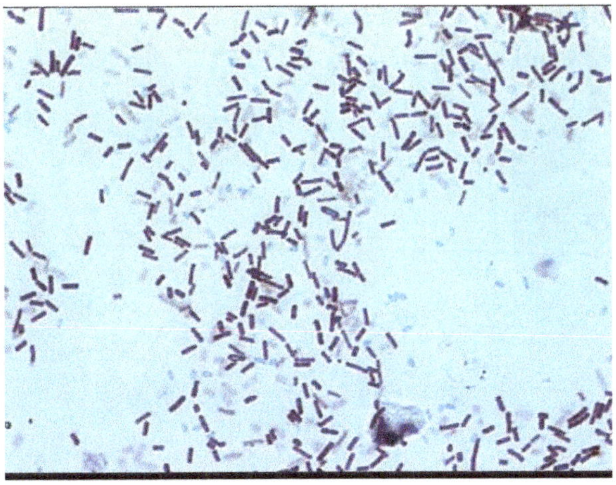

Figure source: *Environmental Leverage*

a. Bacteria
b. Rotifers
c. Stalked ciliates
d. Filaments

228. Identify what is depicted in the image.

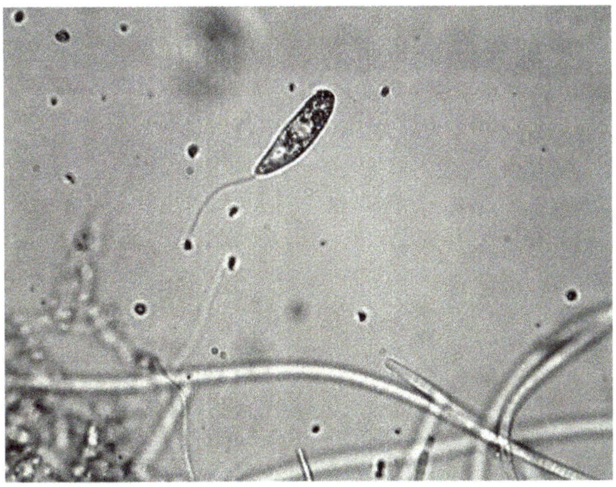

Figure source: *Environmental Leverage*

a. Bacteria
b. Rotifers
c. Stalked ciliates
d. Flagellates

229. Identify what is depicted in the image.

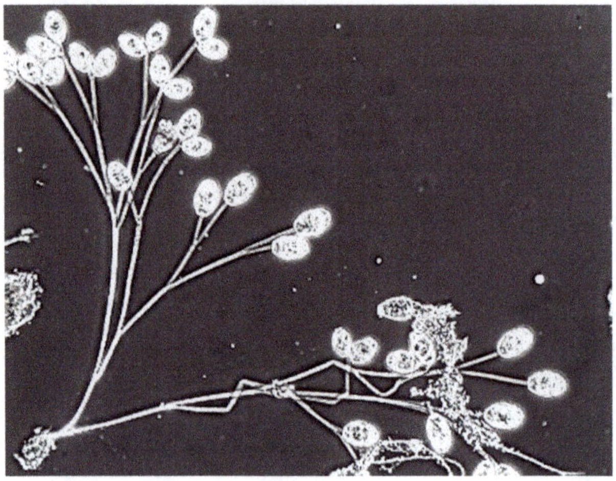

Figure source: *Environmental Leverage*

a. Bacteria
b. Rotifers
c. Stalked ciliates
d. Filaments

230. Identify what is depicted in the image.

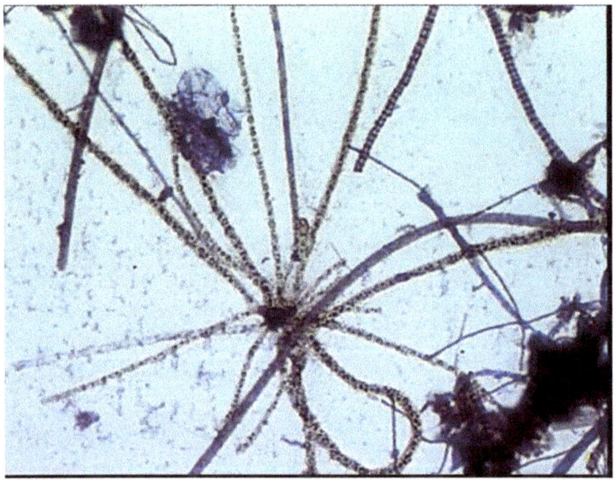

Figure source: *Environmental Leverage*

a. Bacteria
b. Rotifers
c. Stalked ciliates
d. Filaments

231. Identify what is depicted in the image.

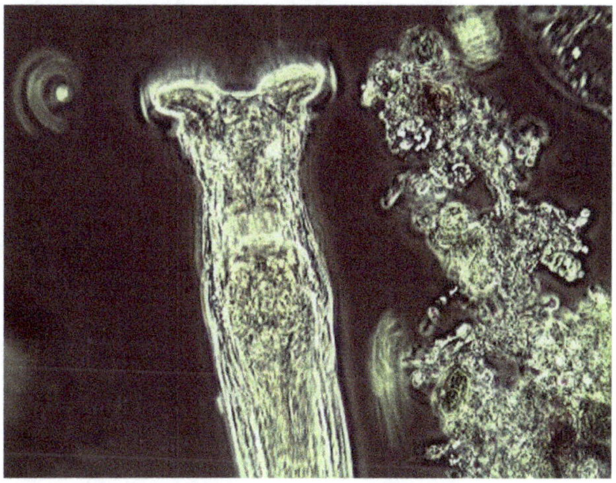

Figure source: *Environmental Leverage*

a. Bacteria
b. Rotifers
c. Stalked ciliates
d. Filaments

This page intentionally blank.

I. Wastewater Collections, Answers

Math

1. **b.** 4.0 ft/s

   Ref.: page 524, Operation and Maintenance of Wastewater Collection Systems, Volume 1, 8th Edition

   Velocity = distance/time
   Convert 5 min to seconds
   5 min × 60 s/1 min = 300 s
   Velocity = 1,200 ft/300 s = 4.0 ft/s

   Wastewater Collections, Introductory (Grades 1 and 2)

   Collection System Operation, Maintenance, & Restoration

2. **d.** 810 ft

   Ref.: page 524, Operation and Maintenance of Wastewater Collection Systems, Volume 1, 8th Edition

   Convert 9 min to seconds
   9 min × 60 s/1 min = 540 s
   Distance = velocity × time = 1.5 ft/1 s × 540 s = 810 ft

   Wastewater Collections, Introductory (Grades 1 and 2)

   Collection System Monitoring, Evaluation, & Adjustment

3. **a.** 3.49 ft^3/s

   Ref.: page 95, Basic Science Concepts and Applications, 4th Edition

   Convert 2.25 mgd to cubic feet per second
   2.25 mil gal/1 day × 1 day/24 h × 1 h/60 min × 1 min/60 s × 1,0000,000 gal/1 mil gal × ft^3/7.48 gal = (2.25)(1,000,000)/(24)(60)(60)(7.48)= 3.49 ft^3/s

   or 2.25 mgd × 1.55 ft^3/s/mgd = 3.49 ft^3/s

   Collection System Monitoring, Evaluation, & Adjustment

   Wastewater Collections, Introductory (Grades 1 and 2)

   Collection System Operation, Maintenance, & Restoration

4. **c.** 8.33 min

Ref.: page 524, Operation and Maintenance of Wastewater Collection Systems, Volume 1, 8th Edition

Time = distance/velocity = 1,000 ft × 1 s/2 ft = 500 s
500 s × 1 min/60 s = 8.33 min

Wastewater Collections, Introductory (Grades 1 and 2)

Collection System Monitoring, Evaluation, & Adjustment

Collection System Operation, Maintenance, & Restoration

5. **c.** 538.56 gpm

Ref.: page 524, Operation and Maintenance of Wastewater Collection Systems, Volume 1, 8th Edition

1.2 ft^3/s × 7.48 gal/ft^3 × 60 s/min = (1.2)(7.48)(60) = 538.56 gpm

Wastewater Collections, Introductory (Grades 1 and 2)

Collection System Monitoring, Evaluation, & Adjustment

Collection System Operation, Maintenance, & Restoration

6. **c.** 345.60 gpm

Ref.: page 526, Operation and Maintenance of Wastewater Collection Systems, Volume 1, 8th Edition

Q = velocity × area
D = 8 in. = 0.67 ft

Area = (π/4)(D^2) = (3.14/4)(0.67)(0.67) = 0.35 ft^2

Q_{gpm} = velocity × area = 2.2 ft/s × 0.35 ft^2 × 60 s/min × 7.48 gal/ft^3 = 345.6 gpm

Wastewater Collections, Advanced (Grades 3 and 4)

Collection System Monitoring, Evaluation, & Adjustment

Collection System Operation, Maintenance, & Restoration

7. **d.** 2.55 ft/s

Ref.: page 526, Operation and Maintenance of Wastewater Collection Systems, Volume 1, 8th Edition

Velocity = *Q*/A
D = 6 in. = 0.5 ft

Area = (3.14/4)(D^2) = (0.785)(0.5)(0.5) = 0.196 ft^2
Velocity = 0.5 ft^3/s × 1/0.196 ft^2 = 2.55 ft/s

Wastewater Collections, Advanced (Grades 3 and 4)

Collection System Monitoring, Evaluation, & Adjustment

Collection System Operation, Maintenance, & Restoration

8. **c.** 10 in.

Ref.: page 526, Operation and Maintenance of Wastewater Collection Systems, Volume 1, 8th Edition

Area = $(3.14/4)(D^2) = 0.785(D^2) = 0.54\ ft^2$
$\sqrt{(0.54\ ft^2/0.785)} = D = 0.83\ ft$

$0.83\ ft \times 12\ in./ft = 9.96\ in.$
$\cong 10.0\ in.$

Wastewater Collections, Advanced (Grades 3 and 4)

Collection System Monitoring, Evaluation, & Adjustment

Collection System Operation, Maintenance, & Restoration

9. **d.** 21.0 cfs

Ref.: page 525, Operation and Maintenance of Wastewater Collection Systems, Volume 1, 8th Edition

Q = velocity × area
Area = width × depth = $6\ ft \times 2\ ft = 12\ ft^2$
$= 1.75\ ft/s \times 12\ ft^2 = 21.0\ ft^3/s$

Wastewater Collections, Advanced (Grades 3 and 4)

Collection System Monitoring, Evaluation, & Adjustment

10. **a.** 3.50 ft

Ref.: page 525, Operation and Maintenance of Wastewater Collection Systems, Volume 1, 8th Edition

$A = Q/V = 39.38\ ft^3/s \times s/2.25\ ft = 39.38\ ft^3/2.25\ ft = 17.50\ ft^2$

A = width × depth = $17.50\ ft^2 = 5\ ft \times$ depth
$17.50\ ft^2/5\ ft = 3.5\ ft$ of depth

Wastewater Collections, Advanced (Grades 3 and 4)

Collection System Monitoring, Evaluation, & Adjustment

11. **a.** 10.83 psi

Ref.: page 218, Basic Science Concepts and Applications, 4th Edition

$P = 25\ ft \times 0.433\ psi/ft = 10.83\ psi$
or $25\ ft \times psi/2.31\ ft = 10.83\ psi$

Wastewater Collections, Introductory (Grades 1 and 2)

Collection System Monitoring, Evaluation, & Adjustment

Equipment Operation, Evaluation, & Maintenance

Lift Station Operation & Maintenance

12. **d.** 779,400.00 lb

Ref.: page 223, Basic Science Concepts and Applications, 4th Edition

Force = pressure × area
Area = 25 ft × 50 ft = 1,250 ft^2
Pressure = 10 ft × psi/2.31 ft = 4.33 psi
Convert psi to lb/ft^2 = 4.33 lb/$in.^2$ × 144 $in.^2/ft^2$ = 623.52 lb/ft^2
Force = pressure × area = 623.52 lb/ft^2 × 1,250 ft^2 = 779,400 lb

Wastewater Collections, Advanced (Grades 3 and 4)

Collection System Monitoring, Evaluation, & Adjustment

Collection System Operation, Maintenance, & Restoration

13. **c.** 34.64 psi

Ref.: page 218, Basic Science Concepts and Applications, 4th Edition

Pressure = 80 ft × psi/2.31 ft = 34.64 psi
or 80 ft × 0.433 psi/ft = 34.64 psi

Wastewater Collections, Advanced (Grades 3 and 4)

Collection System Monitoring, Evaluation, & Adjustment

Collection System Operation, Maintenance, & Restoration

14. **a.** 21.10°C

Ref.: pages 113-116, Basic Science Concepts and Applications, 4th Edition, page 537, Operation and Maintenance of Wastewater Collection Systems, Volume 1, 8th Edition

°C = (°F − 32) × 5/9 = (70 − 32) × 5/9 = (38) × 5/9 = 190/9
= 21.1°C

Wastewater Collections, Introductory (Grades 1 and 2)

Collection System Monitoring, Evaluation, & Adjustment

Collection System Operation, Maintenance, & Restoration

15. **c.** 41.00°F

Ref.: pages 113-116, Basic Science Concepts and Applications, 4th Edition, page 537, Operation and Maintenance of Wastewater Collection Systems, Volume 1, 8th Edition

°F = (°C × 9/5) + 32 = (5 × 9/5) + 32 = 9 + 32 = 41°F

Wastewater Collections, Introductory (Grades 1 and 2)

Collection System Monitoring, Evaluation, & Adjustment

Collection System Operation, Maintenance, & Restoration

16. **d.** 17.76 hp

Ref.: pages 540-543, Operation and Maintenance of Wastewater Collection Systems, Volume 1, 8th Edition

hp = $Q_{gpm} \times H_{ft}/3{,}960 \times E = (750)(75)/(3{,}960)(0.80) = 56{,}250/3{,}168$
$= 17.76$ hp

Wastewater Collections, Advanced (Grades 3 and 4)

Collection System Operation, Maintenance, & Restoration

Equipment Operation, Evaluation, & Maintenance

Lift Station Operation & Maintenance

17. **c.** 3.92 ft^3/s

Ref.: pages 524-525, Operation and Maintenance of Wastewater Collection Systems, Volume 1, 8th Edition

Flow rate = Q = volume/time
Volume = length × width × height = 20 ft × 10 ft × 5 ft = 1,000 ft^3
Time = 4 min, 15 s = 4 min × 60 s/min = 240 + 15 = 255 s
Flow rate = volume/time = 1,000 ft^3/255 s = 3.92 ft^3/s

Wastewater Collections, Advanced (Grades 3 and 4)

Collection System Operation, Maintenance, & Restoration

18. **a.** 0.002

Ref.: pages 62-63, Operation and Maintenance of Wastewater Collection Systems, Volume 1, 5th Edition; pages 52-55, Operation and Maintenance of Wastewater Collection Systems, Volume 1, 8th Edition

Slope = $\Delta y/\Delta x$
Δy = 250.47 ft − 248.95 ft = 1.52 ft
Δx = 750 ft
Slope = $\Delta y/\Delta x$ = 1.52 ft/750 ft = 0.002

Wastewater Collections, Introductory (Grades 1 and 2)

Collection System Operation, Maintenance, & Restoration

19. **d.** 325.39

Ref.: pages 62-63, Operation and Maintenance of Wastewater Collection Systems, Volume 1, 5th Edition; pages 52-55, Operation and Maintenance of Wastewater Collection Systems, Volume 1, 8th Edition

Slope = change in elevation/change in distance = $\Delta y/\Delta x$ = 2 percent or 0.02
0.02 = Δy/500
Δy = (0.02)(500) = 10 ft elevation = 315.39 + 10 = 325.39
Upstream elevation = downstream elevation + Δy

Wastewater Collections, Advanced (Grades 3 and 4)

Collection System Operation, Maintenance, & Restoration

20. **b.** 252.00 lb

Ref.: page 223, Basic Science Concepts and Applications, 4th Edition; page 479, Operation and Maintenance of Wastewater Collection Systems, Volume 1, 8th Edition

$P = 5 \text{ lb/in.}^2 \times 144 \text{ in.}^2/\text{ft}^2 = 720 \text{ lb/ft}^2$
$D = 8 \text{ in.} = 0.67 \text{ ft}$
$A = (3.14/4)(D^2) = (0.785)(0.67)(0.67) = 0.35 \text{ ft}^2$
$F = \text{pressure} \times \text{area} = 720 \text{ lb/ft}^2 \times 0.35 \text{ ft}^2 = 252 \text{ lb}$

Wastewater Collections, Introductory (Grades 1 and 2)

Collection System Monitoring, Evaluation, & Adjustment

Collection System Operation, Maintenance, & Restoration

Equipment Operation, Evaluation, & Maintenance

21. **d.** 24,041 ft^2

Ref.: pages 65-68, Basic Science Concepts and Applications, 4th Edition

$\text{Circumference} = \pi \times D$
$D = \text{circumference}/\pi = 549.5 \text{ ft}/3.14 = 175 \text{ ft}$
$A = (\pi/4)(D^2) = (3.14/4)(175)(175)$
$= (0.785)(175)(175) = 24{,}041 \text{ ft}^2$

Wastewater Treatment, Introductory (Grades 1 and 2)

Collection System Operation, Maintenance, & Restoration

22. **b.** 62 ft

Ref: page 92, Basic Science Concepts and Applications, 4th Edition

Convert pounds of water to cubic feet, assumed as the total volume of the tank

$\text{Volume} = 1{,}500{,}000 \text{ lb} \times \text{ft}^3/62.4 \text{ lb} = 24{,}038.5 \text{ ft}^3$

$\text{Area} = \text{volume/depth} = 24{,}038.5 \text{ ft}^3/8 \text{ ft} = 3{,}005 \text{ ft}^2$

$\text{Area} = (3.14/4)(D^2)$

$D = \sqrt{(\text{Area}/(3.14/4))} = \sqrt{(3{,}005 \text{ ft}^2/0.785)} = \sqrt{3{,}828} = 61.87 \text{ ft}$

Wastewater Collections, Advanced (Grades 3 and 4)

Equipment Operation, Evaluation, & Maintenance

23. **c.** 2.64 h

Ref.: pages 191-193, Basic Science Concepts and Applications, 4th Edition

Detention time = volume/flow rate
Volume = area × depth
Area = (3.14/4)(D^2) = (0.785)(60)(60) = 2,826 ft^2
D = 2r = (2)(30) = 60 ft
Flow rate – 1.5 mgd = 1,500,000 gpd

Volume = area × depth = 2,826 ft^2 × 8 ft = 22,608 ft^3
Volume (gallons) = 22,608 ft^3 × 7.48 gal/ft^3 = 169,108 gal
Detention time = volume/flow rate = 169,108 gal × DAY/1,500,000 gal = 0.11 day × 24 h/1 day = 2.64 h

Wastewater Collections, Introductory (Grades 1 and 2)

Collection System Operation, Maintenance, & Restoration

24. **c.** 1.83 mgd

Ref.: pages 191-193, Basic Science Concepts and Applications, 4th Edition

Flow rate = volume/time
Volume = area × depth
Area = (π/4)(D^2) = (3.14/4)(55)(55) = 2,375 ft^2
Volume = area × depth = 2,375 ft^2 × 7.5 ft = 17,813 ft^3 × 7.48 gal/ft^3 = 133,241 gal
Volume = 133,241 gal/1,000,000 gal = 0.133241 mil gal
Time = 1 h, 45 min = 1.75 h = 1.75 h × 1 day/24 hr = 0.073 day
Flow rate = volume/time = 0.133241 mil gal = 1.83 mgd

Wastewater Collections, Advanced (Grades 3 and 4)

Collection System Operation, Maintenance, & Restoration

25. **b.** 2.77 h

Ref.: pages 191-193, Basic Science Concepts and Applications, 4th Edition

Time = volume/flow rate
Volume = area × depth
Area = (π/4)(D^2) = (0.785)(75)(75) = 4,415.63 ft^2
Volume = area × depth = 4,415.63 ft^2 × 7 = 30,909.41 ft^3
Time = volume/flow rate = 30,909.41 ft^3 × s/3.1 ft^3 = 9,970.78 s
9,970.78 s = 166.18 min = 2.77 h

Wastewater Collections, Advanced (Grades 3 and 4)

Collection System Monitoring, Evaluation, & Adjustment

Collection System Operation, Maintenance, & Restoration

26. **a.** 6.71 percent

Ref.: pages 380-381, Basic Science Concepts and Applications, 4th Edition

By weight: 50 gal × 8.34 lb/gal = 417 lb
Percent strength = weight of solvent/weight of solute × 100 = (30/(30 + 417)) × 100
= 30/447 × 100 = 6.71 percent

Wastewater Collections, Introductory (Grades 1 and 2)

Equipment Operation, Evaluation, & Maintenance

27. **c.** 1.22 mgd

Ref.: pages 57-58, Basic Science Concepts and Applications, 4th Edition

Average daily flows = total flows/number of days
= ((1.25) + (1.29) + (1.18) + (1.20) + (1.22) + (1.28) + (1.10))/7
= 8.52/7 = 1.22 mgd

Wastewater Collections, Introductory (Grades 1 and 2)

Lift Station Operation & Maintenance

28. **b.** 8 sections

Ref. page 572, Operation and Maintenance of Wastewater Collection Systems, Volume 1, 5th Edition

18 feet per section
Sections needed = 137 ft × section/18 ft
= 137/18 = 7.6 sections
≡ 8 sections of pipe

Wastewater Collections, Introductory (Grades 1 and 2)

Collection System Operation, Maintenance, & Restoration

29. **c.** 24.50 ft

Ref. page 573, Operation and Maintenance of Wastewater Collection Systems, Volume 1, 5th Edition

0.035 ft/ft = $\Delta y/\Delta x$
Δx = 700 ft
0.035 = Δy/700
Δy = (0.035)(700)
= 24.5 ft

Wastewater Collections, Introductory (Grades 1 and 2)

Collection System Monitoring, Evaluation, & Adjustment

Collection System Operation, Maintenance, & Restoration

30. **d.** 9.00 mg/L

Ref.: pages 451-453, Basic Science Concepts and Applications, 4th Edition

Chlorine dosage − chlorine demand = chlorine residual
Chlorine dosage − 7.9 mg/L = 1.1 mg/L
Chlorine dosage = chlorine demand + chlorine residual
= 7.9 mg/L + 1.1 mg/L = 9.0 mg/L

Wastewater Collections, Introductory (Grades 1 and 2)

Equipment Operation, Evaluation, & Maintenance

31. **b.** 1,333.33 yd^3

Ref.: page 81, Basic Science and Concepts Applications, 4th Edition

Volume = w × d × L
4 ft × 6 ft × 1,500 ft = 36,000 ft^3
36,000 ft^3 × yd^3/27 ft^3 = 1,333.33 yd^3

Wastewater Collections, Introductory (Grades 1 and 2)

Collection System Operation, Maintenance, & Restoration

32. **c.** 117,436 gal

Ref.: page 81, Basic Science and Concepts Applications, 4th Edition

Volume = Area × Depth
A = $(\pi/4)(D^2)$ = (0.785)(50 ft)(50 ft) = 1,962.5 ft^2
Depth = 8 ft
Volume = 1,962.5 ft^2 × 8 ft = 15,700 ft^3
15,700 ft^3 x 7.48 gal/ft^3 = 117,436 gal

Wastewater Collections, Advanced (Grades 3 and 4)

Equipment Operation, Evaluation, & Maintenance

33. **b.** 57 min

Ref.: page 28, Operation and Maintenance of Wastewater Collection Systems, Volume II, 7th Edition

Known:
Storage volume = 2,000 gal
Pump capacity = 75 gpm
Wet well inflow = 40 gpm

Cycle time = storage volume (gal)/pump capacity (gpm) − wet well inflow (gpm)
= 2,000 gal/(75 − 40) gpm = 57.14 min

Wastewater Collections, Advanced (Grades 3 and 4)

Collection System Operation, Maintenance, & Restoration

Lift Station Operation & Maintenance

34. **c.** 1,485 gpm

Ref.: page 28, Operation and Maintenance of Wastewater Collection Systems, Volume II, 7th Edition

Known:
Cycle time = 45 min
Pump capacity = 65 gpm
Wet well inflow = 32 gpm

Storage volume = (cycle time) (pump capacity – inflow rate)
= (45 min)(65 – 32)
= (45)(33)
= 1,485 gal

Wastewater Collections, Advanced (Grades 3 and 4)

Collection System Operation, Maintenance, & Restoration

Lift Station Operation & Maintenance

35. **a.** 96

Ref.: page 587, Operation and Maintenance of Wastewater Collection Systems, Volume II, 7th Edition

Performance indicator = (costs, $/mile)(stoppages, number/mile)
Costs = annual expenditures/length of sewer system = $2,750,000/2,000 = $1,375/mile
Stoppages = number of stoppages/length of sewer system = 145/2,000 = 0.07
Performance indicator = ($1,375/mile)(0.07) = 96.25

Wastewater Collections, Advanced (Grades 3 and 4)

Security, Safety, & Administrative Procedures

Collection System Monitoring, Evaluation, & Adjustment

36. **d.** 72 percent

Ref.: page 587, Operation and Maintenance of Wastewater Collection Systems, Volume II, 7th Edition

Performance rating = (standard)(100%)/performance indicator = (90)(100)/125
= 72 percent

Wastewater Collections, Advanced (Grades 3 and 4)

Security, Safety, & Administrative Procedures

37. **c.** 502.4 lb

Ref.: page 113, Operation and Maintenance of Wastewater Collection Systems, Volume I, 5th Edition

Force = pressure × area
Pressure = 10 psi
Area = $(\pi/4)(D^2)$ = (0.785)(8)(8) = 50.24 in.^{2}
Force = (10 psi)(50.24 in.^{2}) = 502.4 lb

Wastewater Collections, Advanced (Grades 3 and 4)

Collection System Operation, Maintenance, & Restoration

38. **b.** 904 lb

Ref.: page 113, Operation and Maintenance of Wastewater Collection Systems, Volume I, 5th Edition

Force = pressure × area
Pressure = 8 psi
Area = $(\pi/4)(D^2) = (0.785)(12)(12) = 113.04 \text{ in.}^2$
Force = $8 \text{ lb/in.}^2 \times 113.04 \text{ in.}^2 = 904.32$ lb

Wastewater Collections, Advanced (Grades 3 and 4)

Collection System Monitoring, Evaluation, & Adjustment

Collection System Operation, Maintenance, & Restoration

39. **d.** 1,566 gpd

Ref.: page 117, Operation and Maintenance of Wastewater Collection Systems, Volume I, 5th Edition

Volume, gallons = $(0.785)(5)(5)(8/12)(7.48 \text{ gal/ft}^3) = 97.86$ gal
(97.86 gal)(24 h/day)/1.5 h = 1,566 gpd

Wastewater Collections, Advanced (Grades 3 and 4)

Collection System Monitoring, Evaluation, & Adjustment

Collection System Operation, Maintenance, & Restoration

40. **a.** 87.50 percent

Ref.: page 550, Operation and Maintenance of Wastewater Collection Systems, Volume I, 5th Edition

Influent BOD = 240 mg/L
Effluent BOD = 30 mg/L
Plant BOD efficiency = ((in − out)/in)(100%) = ((240 − 30)/240)(100)= 87.5 percent

Wastewater Collections, Introductory (Grades 1 and 2)

Security, Safety, & Administrative Procedures

Collection System Monitoring, Evaluation, & Adjustment

41. **c.** 40.5 mg/L

Ref.: page 550, Operation and Maintenance of Wastewater Collection Systems, Volume I, 5th Edition

Plant BOD efficiency = ((in − out)/in)(100)
Influent BOD = 270 mg/L
Removal efficiency 85 percent
Plant BOD efficiency = 85 = ((270 − x)/270)(100)
(85)(270)/100 = 270 − x
229.5 = 270 − x
x = 40.5 mg/L

Wastewater Collections, Introductory (Grades 1 and 2)

Security, Safety, & Administrative Procedures

Collection System Monitoring, Evaluation, & Adjustment

42. **b.** 1.78 mg/L

Ref.: pages 450-452, Basic Science Concepts and Applications, 4th Edition

Chlorine feed rate (lb/day) = chlorine dosage (mg/L) × flow (mgd) × 8.34 lb/day
700 gal/min × mil gal/1,000,000 gal × 60 min/1 h × 24 h/day = 1.01 mgd
Chlorine dosage (mg/L) = chlorine feed rate/flow (mgd) 8.34 lb = 15/(1.01)(8.34)
= 1.78 mg/L

Wastewater Collections, Introductory (Grades 1 and 2)

Equipment Operation, Evaluation, & Maintenance

43. **a.** 16.81 lb/day

Ref.: pages 450-452, Basic Science Concepts and Applications, 4th Edition

Chlorine feed rate (lb/day) = chlorine dosage (mg/L) × flow (mgd) × 8.34
500 gal/min × mil gal/1,000,000 gal × 60 min/1 h × 24 h/day = 0.72 mgd
Chlorine feed rate (lb/day) = (2.8 mg/L)(0.72 mgd)(8.34)
= 16.81 lb/day

Wastewater Collections, Introductory (Grades 1 and 2)

Equipment Operation, Evaluation, & Maintenance

44. **d.** 8.25 ft^3/s

Ref.: pages 305-308, Basic Science Concepts and Applications, 4th Edition

Q = velocity × area= 2.2 ft/s × 2.5 ft × 1.5 ft= 8.25 ft^3/s

Wastewater Collections, Advanced (Grades 3 and 4)

Collection System Monitoring, Evaluation, & Adjustment

Collection System Operation, Maintenance, & Restoration

45. **b.** 2.22 ft/s

Ref.: pages 305-308, Basic Science Concepts and Applications, 4th Edition

Velocity = flow/area
Flow = 1.2 ft^3/s
Area = $(\pi/4)(D^2)$ = (0.785)(10/12)(10/12) = 0.54 ft^2
Velocity = (1.2 ft^3/s)/0.54 ft^2 = 2.22 ft/s

Wastewater Collections, Advanced (Grades 3 and 4)

Collection System Monitoring, Evaluation, & Adjustment

46. **c.** 28.47 hp

Ref.: pages 68-73, Operation and Maintenance of Wastewater Collection Systems, Volume II, 7th Edition

HP = flow (gpm) × H (ft)/3,960 × E
Flow = 2.3 mil gal/day × 1 day/24 h × 1,000,000 gal/mil gal × 1 h/60 min = 1,597.2 gpm
= (1,597.2 gpm)(60 ft)/(3,960)(0.85)
= 28.47 hp

Wastewater Collections, Advanced (Grades 3 and 4)

Collection System Operation, Maintenance, & Restoration

Lift Station Operation & Maintenance

47. **a.** 2.05 mgd

Ref.: pages 68-73, Operation and Maintenance of Wastewater Collection Systems, Volume II, 7th Edition

HP = (flow (gpm) × H (ft))/(3,960 × E)
30 = (flow (gpm) × 75)/(3,960 × 0.90)
((30)(3,960)(0.90))/75 = flow (gpm)
= 1,425.6 gpm
1,425.6 gal/min × mil gal/1,000,000 gal × 60 min/1 h × 24 h/day = 2.05 mgd

Wastewater Collections, Advanced (Grades 3 and 4)

Collection System Monitoring, Evaluation, & Adjustment

Collection System Operation, Maintenance, & Restoration

Lift Station Operation & Maintenance

48. **d.** 34,515 lb

Ref.: pages 206-210, Basic Science Concepts and Applications, 4th Edition

(8.34 lb/gal)(0.92) = 7.67 lb/gal
(4,500 gal)(7.67 lb/gal) = 34,515 lb

Wastewater Collections, Introductory (Grades 1 and 2)

Security, Safety, & Administrative Procedures

Equipment Operation, Evaluation, & Maintenance

49. **b.** 0.89

Ref.: pages 206-210, Basic Science Concepts and Applications, 4th Edition

Weight (lb) = Volume (gal) × (8.34 lb/gal)(specific gravity)
40,000 lb = 5,391 gal × 8.34 × specific gravity
40,000 lb/(5,391 gal × 8.34) = specific gravity = 0.89

Wastewater Collections, Introductory (Grades 1 and 2)

Collection System Monitoring, Evaluation, & Adjustment

50. **a.** 7.48 gpm

Ref.: pages 305-308, Basic Science Concepts and Applications, 4th Edition

Volume = 10 ft × 6 ft × 2 ft = 120 ft^3 × 7.48 gal/ft^3 = 898 gal
2 h × 60 min/1 h = 120 min
Pumping rate (gpm) = 898 gal/120 min = 7.48 gpm

Wastewater Collections, Advanced (Grades 3 and 4)

Collection System Monitoring, Evaluation, & Adjustment

Equipment Operation, Evaluation, & Maintenance

51. **c.** 11.15 mgd

Ref.: page 308, Basic Science Concepts and Applications, 4th Edition

Flow = velocity × area
Width = 5 ft
Depth = 18 in. = 1.5 ft
= 2.3 ft/s × 5 × 1.5 = 17.25 ft^3/s
17.25 ft^3/s × 7.48 gal/ft^3 × 60 s/1 min × 60 min/1 h × 24 h/day × mil gal/1,000,000 gal = 11.15 mgd

Wastewater Collections, Advanced (Grades 3 and 4)

Collection System Monitoring, Evaluation, & Adjustment

52. **d.** 4.76 ft

Ref.: page 310, Basic Science Concepts and Applications, 4th Edition

Flow =
8 mil gal/day × 1,000,000 gal/mil gal × ft^3/7.48 gal × 1 day/24 h × 1 h/60 min × 1 min/60 s = 12.38 ft^3/s
Flow = velocity × area
Area = flow/velocity = 12.38 ft^3/1.3 ft/s = 9.52 ft^2
Width = area/depth = 9.52 ft/2.0 ft = 4.76 ft

Wastewater Collections, Advanced (Grades 3 and 4)

Collection System Monitoring, Evaluation, & Adjustment

53. **c.** 178 trucks

Ref.: page 81, Basic Science Concepts and Applications, 4th Edition

Volume = 4 ft × 10 ft × 1,200 ft = 48,000 ft^3
48,000 ft^3 × yd^3/27 ft^3 = 1,777.78 yd^3
If each dump truck holds 10 yd^3, the number of 10 yd^3 trucks will be
1,777.78 yd^3 × truck/10 yd^3 = 177.78 ≅ 178 trucks

Wastewater Collections, Introductory (Grades 1 and 2)

Security, Safety, & Administrative Procedures

Collection System Operation, Maintenance, & Restoration

54. **b.** 7.02 ft

Ref.: page 81, Basic Science Concepts and Applications, 4th Edition

$1{,}300\ yd^3 \times 27\ ft^3/yd^3 = 35{,}100\ ft^3$
Volume = W × L × D
$35{,}100\ ft^3 = 5\ ft \times 1{,}000\ ft \times D$
$35{,}100\ ft^3/(5\ ft \times 1{,}000) = D = 7.02\ ft$

Wastewater Collections, Advanced (Grades 3 and 4)

Wastewater Collections, Introductory (Grades 1 and 2)

Collection System Operation, Maintenance, & Restoration

55. **b.** 9.54 ft

Ref: page 81, Basic Science Concepts and Applications, 4th Edition

$V = 35{,}000\ gal \times ft^3/7.48\ gal = 4{,}679.14\ ft^3$
$A = (\pi/4)(D^2) = (0.785)(25\ ft)(25\ ft) = 490.63\ ft^2$
$Depth = volume/area = 4{,}679.14\ ft^3/490.63\ ft^2 = 9.54\ ft$

Wastewater Collections, Introductory (Grades 1 and 2)

Equipment Operation, Evaluation, & Maintenance

General

1. **c.** clear wells

Ref: page 2, Section 1.1.1, What is a Wastewater Collection System, O & M of Wastewater Collection Systems, Volume 1, 8th edition

Wastewater Collections, Introductory (Grades 1 and 2)

Collection System Operation, Maintenance, & Restoration

2. **a.** Blockages in the sewer mains, businesses, or homes

Ref.: pages 2-3, Section 1.1.2, Why is Operation and Maintenance of Wastewater Collection Systems Important, O & M of Wastewater Collection Systems, Volume 1, 8th edition

Wastewater Collections, Introductory (Grades 1 and 2)

Collection System Operation, Maintenance, & Restoration

3. **d.** All of the above

Ref.: page 3, Section 1.1.2, Why is Operation and Maintenance of Wastewater Collection Systems Important, O & M of Wastewater Collection Systems, Volume 1, 8th edition

Wastewater Collections, Introductory (Grades 1 and 2)

Collection System Monitoring, Evaluation, & Adjustment

Collection System Operation, Maintenance, & Restoration

Equipment Operation, Evaluation, & Maintenance

4. **c.** Domestic, regional, and state services

   Ref.: page 9, Section 1.2.1, Overview of O & M Challenges, O & M of Wastewater Collection Systems, Volume 1, 8th edition

   Wastewater Collections, Introductory (Grades 1 and 2)

   Collection System Operation, Maintenance, & Restoration

5. **b.** Operate and maintain the collection system to convey wastewater to the plant

   Ref.: page 5, Section 1.1.3, Working as a Wastewater Collection System Operator, O & M of Wastewater Collection Systems, Volume 1, 8th edition

   Wastewater Collections, Introductory (Grades 1 and 2)

   Collection System Operation, Maintenance, & Restoration

   Equipment Operation, Evaluation, & Maintenance

6. **b.** Proper bedding around the sewer mains

   Ref.: page 9, Section 1.2.1 Overview of O & M Challenges, O & M of Wastewater Collection Systems, Volume 1, 8th edition

   Wastewater Collections, Introductory (Grades 1 and 2)

   Collection System Operation, Maintenance, & Restoration

7. **c.** Infiltration and exfiltration

   Ref.: page 9, Section 1.2.1, Overview of O & M Challenges, O & M of Wastewater Collection Systems, Volume 1, 8th edition

   Wastewater Collections, Introductory (Grades 1 and 2)

   Collection System Monitoring, Evaluation, & Adjustment

   Collection System Operation, Maintenance, & Restoration

8. **a.** Chlorine

   Ref.: page 11, Section 1.2.2, Why Wastewater Collection Systems were built, 1st paragraph, O & M of Wastewater Collection Systems, Volume 1, 8th edition

   Wastewater Collections, Introductory (Grades 1 and 2)

   Security, Safety, & Administrative Procedures

   Collection System Monitoring, Evaluation, & Adjustment

   Collection System Operation, Maintenance, & Restoration

9. **c.** pipe failure and/or settlement

Ref.: page 12, Section 1.2.4, Proposed SSO Rule and CMOM Program, O & M of Wastewater Collection Systems, Volume 1, 8th edition

Wastewater Collections, Introductory (Grades 1 and 2)

Collection System Monitoring, Evaluation, & Adjustment

Collection System Operation, Maintenance, & Restoration

10. **b.** A survey to be completed by the customer

Ref.: page 13, Section 1.2.4, Proposed SSO Rule and CMOM Program, O & M of Wastewater Collection Systems, Volume 1, 8th edition

Wastewater Collections, Introductory (Grades 1 and 2)

Security, Safety, & Administrative Procedures

Collection System Operation, Maintenance, & Restoration

11. **c.** The turbidity of the water at the water treatment plant intake

Ref.: page 13, in the box at the bottom right corner, O & M of Wastewater Collection Systems, Volume 1, 8th edition

Wastewater Collections, Advanced (Grades 3 and 4)

Wastewater Collections, Introductory (Grades 1 and 2)

Security, Safety, & Administrative Procedures

Collection System Monitoring, Evaluation, & Adjustment

Collection System Operation, Maintenance, & Restoration

12. **b.** Lift the wastewater to a higher elevation

Ref.: page 22, Section 2.2, Purpose, 2nd paragraph, O & M of Wastewater Collection Systems, Volume 1, 8th edition

Wastewater Collections, Introductory (Grades 1 and 2)

Collection System Operation, Maintenance, & Restoration

Lift Station Operation & Maintenance

13. **a.** The amount of source water treated and distributed

Ref.: page 23, Section 2.3 Quantity of Wastewater, O & M of Wastewater Collection Systems, Volume 1, 8th edition

Wastewater Collections, Introductory (Grades 1 and 2)

Collection System Monitoring, Evaluation, & Adjustment

Collection System Operation, Maintenance, & Restoration

14. **d.** Booster pump stations

    Ref.: pages 33-34, Section 2.4.1 Gravity Collection Systems, O & M of Wastewater Collection Systems, Volume 1, 8th edition

    Wastewater Collections, Introductory (Grades 1 and 2)

    Collection System Operation, Maintenance, & Restoration

    Equipment Operation, Evaluation, & Maintenance

    Lift Station Operation & Maintenance

15. **b.** The topography and ground conditions are not suitable for conventional gravity system

    Ref.: pages 35-36, Section 2.4.2 Low Pressure Collection System, O & M of Wastewater Collection Systems, Volume 1, 8th edition

    Wastewater Collections, Advanced (Grades 3 and 4)

    Wastewater Collections, Introductory (Grades 1 and 2)

    Collection System Monitoring, Evaluation, & Adjustment

    Collection System Operation, Maintenance, & Restoration

16. **a.** Gravity sewers, holding tanks, grinder pumps, and force mains

    Ref.: page 36, Section 2.4.2 Low Pressure Collection System, O & M of Wastewater Collection Systems, Volume 1, 8th edition

    Wastewater Collections, Advanced (Grades 3 and 4)

    Wastewater Collections, Introductory (Grades 1 and 2)

    Collection System Operation, Maintenance, & Restoration

    Equipment Operation, Evaluation, & Maintenance

17. **c.** Gravity sewers, interface unit, and pumps

    Ref.: page 36, Section 2.4.3 Vacuum Collection System, O & M of Wastewater Collection Systems, Volume 1, 8th edition

    Wastewater Collections, Advanced (Grades 3 and 4)

    Wastewater Collections, Introductory (Grades 1 and 2)

    Collection System Operation, Maintenance, & Restoration

    Equipment Operation, Evaluation, & Maintenance

18. **d.** Steel pipe

    Ref.: page 61, Section 2.5.7 Pipe Materials, middle of the page, O & M of Wastewater Collection Systems, Volume 1, 8th edition

    Wastewater Collections, Introductory (Grades 1 and 2)

    Collection System Operation, Maintenance, & Restoration

19. **c.** Epoxy joints

Ref.: page 68, Section 2.5.8 Pipe Joints, O & M of Wastewater Collection Systems, Volume 1, 8th edition

Wastewater Collections, Introductory (Grades 1 and 2)

Collection System Operation, Maintenance, & Restoration

20. **d.** Manholes used in areas with a high groundwater or corrosion are often made with and/or lined with fiberglass

Ref.: page 71, Section 2.5.9 Manhole Materials, O & M of Wastewater Collection Systems, Volume 1, 8th edition

Wastewater Collections, Introductory (Grades 1 and 2)

Collection System Operation, Maintenance, & Restoration

21. **b.** Buffer zones, exclusion zones, and traffic zone

Ref.: page 135, Section 3.4.1 Confined Spaces, 1st paragraph, O & M of Wastewater Collection Systems, Volume 1, 8th edition

Wastewater Collections, Introductory (Grades 1 and 2)

Security, Safety, & Administrative Procedures

Collection System Operation, Maintenance, & Restoration

22. **c.** Flammable gas, vapor, and combustible dust

Ref.: page 137, Section 3.4.2 Atmospheric Hazards, bullet points, O & M of Wastewater Collection Systems, Volume 1, 8th edition

Wastewater Collections, Introductory (Grades 1 and 2)

Security, Safety, & Administrative Procedures

Collection System Monitoring, Evaluation, & Adjustment

Collection System Operation, Maintenance, & Restoration

Equipment Operation, Evaluation, & Maintenance

Lift Station Operation & Maintenance

23. **d.** A periodic assessment of the workers condition and the air should be performed

Ref.: page 175, Section 3.7.3 While Operator is in the Manhole, O & M of Wastewater Collection Systems, Volume 1, 8th edition

Wastewater Collections, Advanced (Grades 3 and 4)

Wastewater Collections, Introductory (Grades 1 and 2)

Security, Safety, & Administrative Procedures

Collection System Operation, Maintenance, & Restoration

24. **a.** The edges of all surface cuts do not have to be smooth and perpendicular when excavating

    Ref.: page 177, Section 3.8 Excavating and Shoring, O & M of Wastewater Collection Systems, Volume 1, 8th edition

    Wastewater Collections, Advanced (Grades 3 and 4)

    Wastewater Collections, Introductory (Grades 1 and 2)

    Security, Safety, & Administrative Procedures

    Collection System Operation, Maintenance, & Restoration

25. **c.** All types are not designed to prevent cave-ins

    Ref.: pages 181-182, Section 3.8.2 Shoring, O & M of Wastewater Collection Systems, Volume 1, 8th edition

    Wastewater Collections, Advanced (Grades 3 and 4)

    Wastewater Collections, Introductory (Grades 1 and 2)

    Security, Safety, & Administrative Procedures

    Collection System Operation, Maintenance, & Restoration

26. **a.** Spread shoring

    Ref.: pages 185-189, Section 3.8.2.5 Types of Shoring, O & M of Wastewater Collection Systems, Volume 1, 8th edition

    Wastewater Collections, Advanced (Grades 3 and 4)

    Wastewater Collections, Introductory (Grades 1 and 2)

    Security, Safety, & Administrative Procedures

    Collection System Operation, Maintenance, & Restoration

27. **d.** The ladder should extend at least 18 in. above the trench

    Ref.: pages 189-190, Section 3.8.2.6 Avoiding Trench Cave-Ins, O & M of Wastewater Collection Systems, Volume 1, 8th edition

    Wastewater Collections, Advanced (Grades 3 and 4)

    Wastewater Collections, Introductory (Grades 1 and 2)

    Security, Safety, & Administrative Procedures

    Collection System Operation, Maintenance, & Restoration

28. **c.** Isolate the equipment or machine and restore energy to it during repairs

    Ref.: page 199, Section 3.9.4.1 Sequence of Lock Out/Tag Out System Procedures, O & M of Wastewater Collection Systems, Volume 1, 8th edition

    Wastewater Collections, Introductory (Grades 1 and 2)

    Security, Safety, & Administrative Procedures

    Equipment Operation, Evaluation, & Maintenance

29. **a.** Biodegradable dilutions

Ref.: page 206, Section 3.12 Hazard Communication Standard and the Worker Right to Know Laws, O & M of Wastewater Collection Systems, Volume 1, 8th edition

Wastewater Collections, Introductory (Grades 1 and 2)

Security, Safety, & Administrative Procedures

Collection System Operation, Maintenance, & Restoration

30. **d.** Bacteria, protozoa, helminths, and viruses

Ref.: page 208, Section 3.12 Hazard Communication Standard and the Worker Right to Know Laws, O & M of Wastewater Collection Systems, Volume 1, 8th edition

Wastewater Collections, Introductory (Grades 1 and 2)

Security, Safety, & Administrative Procedures

31. **b.** are not intended to be used as part of the new construction acceptance process

Ref.: page 224, Section 4.1 Reasons for Inspecting and Testing, O & M of Wastewater Collection Systems, Volume 1, 8th edition

Wastewater Collections, Introductory (Grades 1 and 2)

Security, Safety, & Administrative Procedures

Collection System Operation, Maintenance, & Restoration

32. **c.** Outside forces, events, or natural disasters

Ref.: page 224, Section 4.2.1 Types of Collection System Problems, O & M of Wastewater Collection Systems, Volume 1, 8th edition

Wastewater Collections, Introductory (Grades 1 and 2)

Collection System Monitoring, Evaluation, & Adjustment

Collection System Operation, Maintenance, & Restoration

33. **a.** They are not recognized as a major cause of operating problems

Ref.: page 226, Section 4.2.2 Infiltration and Inflow, O & M of Wastewater Collection Systems, Volume 1, 8th edition

Wastewater Collections, Introductory (Grades 1 and 2)

Collection System Operation, Maintenance, & Restoration

34. **c.** Manholes can perform their job effectively without the attention given to other assets

Ref.: page 229, Section 4.3 Manhole Inspections, O & M of Wastewater Collection Systems, Volume 1, 8th edition

Wastewater Collections, Introductory (Grades 1 and 2)

Security, Safety, & Administrative Procedures

Collection System Operation, Maintenance, & Restoration

35. **c.** An uncommon approach to inspecting sewer mains

    Ref.: page 235, Section 4.4 Closed Circuit Television Inspections, 1st paragraph, O & M of Wastewater Collection Systems, Volume 1, 8th edition

    Wastewater Collections, Introductory (Grades 1 and 2)

    Collection System Operation, Maintenance, & Restoration

36. **b.** The most accurate rate of flow

    Ref.: page 235, Section 4.4 Closed-Circuit Television Inspections, 3rd paragraph, O & M of Wastewater Collection Systems, Volume 1, 8th edition

    Wastewater Collections, Introductory (Grades 1 and 2)

    Collection System Operation, Maintenance, & Restoration

37. **c.** Equipment costs are often more expensive than other inspection processes

    Ref.: page 281, Section 4.5 Smoke Testing, 1st paragraph, O & M of Wastewater Collection Systems, Volume 1, 8th edition

    Wastewater Collections, Introductory (Grades 1 and 2)

    Equipment Operation, Evaluation, & Maintenance

38. **a.** Does not reduce the possibility of dangerous gases getting into a building

    Ref.: page 281, Section 4.5 Smoke Testing, bullet points, O & M of Wastewater Collection Systems, Volume 1, 8th edition

    Wastewater Collections, Introductory (Grades 1 and 2)

    Equipment Operation, Evaluation, & Maintenance

39. **b.** The identification of the types of pipe material in the system

    Ref.: page 282, Section 4.5 Smoke Testing, top of page, O & M of Wastewater Collection Systems, Volume 1, 8th edition

    Wastewater Collections, Introductory (Grades 1 and 2)

    Equipment Operation, Evaluation, & Maintenance

40. **a.** Increase the rate of sewer flow into a building

    Ref.: page 292, Section 4.6 Dye Testing, 1st paragraph, O & M of Wastewater Collection Systems, Volume 1, 8th edition

    Wastewater Collections, Introductory (Grades 1 and 2)

    Equipment Operation, Evaluation, & Maintenance

41. **b.** Should be used exclusively instead of CCTV to inspect sewer mains

    Ref.: page 292, Section 4.6 Dye Testing, bullet points, O & M of Wastewater Collection Systems, Volume 1, 8th edition

    Wastewater Collections, Introductory (Grades 1 and 2)

    Equipment Operation, Evaluation, & Maintenance

42. **c.** All stoppages result in the flow eventually discontinuing

Ref.: page 303, Section 5.1.1 Identifying Problems, 1st paragraph, O & M of Wastewater Collection Systems, Volume 1, 8th edition

Wastewater Collections, Advanced (Grades 3 and 4)

Wastewater Collections, Introductory (Grades 1 and 2)

Collection System Monitoring, Evaluation, & Adjustment

Collection System Operation, Maintenance, & Restoration

43. **a.** Water flowing into the system above 95°F

Ref.: page 303, Section 5.1.1.1 Types of Stoppages, 1st paragraph, O & M of Wastewater Collection Systems, Volume 1, 8th edition

Wastewater Collections, Introductory (Grades 1 and 2)

Collection System Monitoring, Evaluation, & Adjustment

Collection System Operation, Maintenance, & Restoration

44. **d.** The energy used to pump the sewerage

Ref.: page 305, Section 5.1.2 Methods of Cleaning and Maintaining Sewers, 1st paragraph, O & M of Wastewater Collection Systems, Volume 1, 8th edition

Wastewater Collections, Introductory (Grades 1 and 2)

Collection System Monitoring, Evaluation, & Adjustment

Collection System Operation, Maintenance, & Restoration

45. **a.** they cannot remove grit, grease, and debris from the sewer

Ref.: page 305, Section 5.1.2 Methods of Cleaning and Maintaining Sewers, 2nd paragraph, O & M of Wastewater Collection Systems, Volume 1, 8th edition

Wastewater Collections, Introductory (Grades 1 and 2)

Collection System Monitoring, Evaluation, & Adjustment

Collection System Operation, Maintenance, & Restoration

Equipment Operation, Evaluation, & Maintenance

Lift Station Operation & Maintenance

46. **d.** All of the above

Ref.: page 305, Section 5.1.2 Methods of Cleaning and Maintaining Sewers, 3rd paragraph, O & M of Wastewater Collection Systems, Volume 1, 8th edition

Wastewater Collections, Introductory (Grades 1 and 2)

Collection System Operation, Maintenance, & Restoration

Equipment Operation, Evaluation, & Maintenance

47. **c.** Cheaters, extensions, and slings

    Ref.: page 305, Section 5.1.2 Methods of Cleaning and Maintaining Sewers, 4th paragraph, O & M of Wastewater Collection Systems, Volume 1, 8th edition

    Wastewater Collections, Introductory (Grades 1 and 2)

    Equipment Operation, Evaluation, & Maintenance

48. **c.** High velocity cleaners, hand rodders, and power rodders

    Ref.: page 306, Section 5.14 Emergency Cleaning, Table 5.1, O & M of Wastewater Collection Systems, Volume 1, 8th edition

    Wastewater Collections, Introductory (Grades 1 and 2)

    Collection System Operation, Maintenance, & Restoration

49. **b.** High velocity cleaners, power rodders, and bacteria

    Ref.: page 306, Section 5.14 Emergency Cleaning, Table 5.1, O & M of Wastewater Collection Systems, Volume 1, 8th edition

    Wastewater Collections, Advanced (Grades 3 and 4)

    Wastewater Collections, Introductory (Grades 1 and 2)

    Collection System Operation, Maintenance, & Restoration

50. **a.** Chemicals, high velocity cleaners, and power rodders

    Ref.: page 306, Section 5.14 Emergency Cleaning, Table 5.1, O & M of Wastewater Collection Systems, Volume 1, 8th edition

    Wastewater Treatment, Advanced (Grades 3 and 4)

    Wastewater Collections, Introductory (Grades 1 and 2)

    Collection System Operation, Maintenance, & Restoration

51. **a.** High velocities in sewers near restaurants

    Ref.: page 307, Figure 5.1, O & M of Wastewater Collection Systems, Volume 1, 8th edition

    Wastewater Collections, Introductory (Grades 1 and 2)

    Collection System Monitoring, Evaluation, & Adjustment

    Collection System Operation, Maintenance, & Restoration

52. **d.** All of the above

    Ref.: page 307, Figure 5.1, O & M of Wastewater Collection Systems, Volume 1, 8th edition

    Wastewater Collections, Introductory (Grades 1 and 2)

    Collection System Monitoring, Evaluation, & Adjustment

    Collection System Operation, Maintenance, & Restoration

53. **b.** The accumulation of sand and silt

Ref.: page 307, Figure 5.1, O & M of Wastewater Collection Systems, Volume 1, 8th edition

Wastewater Collections, Introductory (Grades 1 and 2)

Security, Safety, & Administrative Procedures

Collection System Monitoring, Evaluation, & Adjustment

Collection System Operation, Maintenance, & Restoration

54. **c.** The distance to the lift station

Ref.: page 315, Section 5.2.1.2 Selecting a Nozzle, O & M of Wastewater Collection Systems, Volume 1, 8th edition

Wastewater Collections, Introductory (Grades 1 and 2)

Equipment Operation, Evaluation, & Maintenance

Lift Station Operation & Maintenance

55. **c.** Low-speed reverse nozzles

Ref.: page 316, 1st paragraph, O & M of Wastewater Collection Systems, Volume 1, 8th edition

Wastewater Collections, Introductory (Grades 1 and 2)

Collection System Operation, Maintenance, & Restoration

Equipment Operation, Evaluation, & Maintenance

Lift Station Operation & Maintenance

56. **b.** Rotational or spinning nozzles

Ref.: page 316, 1st paragraph, O & M of Wastewater Collection Systems, Volume 1, 8th edition

Wastewater Collections, Introductory (Grades 1 and 2)

Collection System Operation, Maintenance, & Restoration

Equipment Operation, Evaluation, & Maintenance

57. **c.** Bags, kites, and poly pigs

Ref.: page 323, Section 5.2.2 Kites, Bags, and Poly pigs, O & M of Wastewater Collection Systems, Volume 1, 8th edition

Wastewater Collections, Introductory (Grades 1 and 2)

Collection System Operation, Maintenance, & Restoration

Equipment Operation, Evaluation, & Maintenance

58. **d.** They are not used for clearing stoppages

Ref.: page 332, Section 5.3.1 Power Rodders, O & M of Wastewater Collection Systems, Volume 1, 8th edition

Wastewater Collections, Introductory (Grades 1 and 2)

Collection System Operation, Maintenance, & Restoration

Equipment Operation, Evaluation, & Maintenance

59. **a.** They can be changed out without replacing the whole reel

Ref.: page 334, Differences between sectional and continuous rods, O & M of Wastewater Collection Systems, Volume 1, 8th edition

Wastewater Collections, Introductory (Grades 1 and 2)

Collection System Operation, Maintenance, & Restoration

Equipment Operation, Evaluation, & Maintenance

60. **c.** Couplings on the sectional rods seldom cause wear on the guide hose liner

Ref.: page 334-335, Differences between sectional and continuous rods, O & M of Wastewater Collection Systems, Volume 1, 8th edition

Wastewater Collections, Introductory (Grades 1 and 2)

Collection System Operation, Maintenance, & Restoration

Equipment Operation, Evaluation, & Maintenance

61. **a.** Porcupine

Ref.: page 336, Figure 5.11, O & M of Wastewater Collection Systems, Volume 1, 8th edition

Wastewater Collections, Introductory (Grades 1 and 2)

Collection System Operation, Maintenance, & Restoration

Equipment Operation, Evaluation, & Maintenance

62. **c.** Square bar corkscrew

Ref.: page 336, Figure 5.11, O & M of Wastewater Collection Systems, Volume 1, 8th edition

Wastewater Collections, Introductory (Grades 1 and 2)

Collection System Operation, Maintenance, & Restoration

Equipment Operation, Evaluation, & Maintenance

63. **b.** Spring blade cutter

Ref.: page 336, Figure 5.11, O & M of Wastewater Collection Systems, Volume 1, 8th edition

Wastewater Collections, Introductory (Grades 1 and 2)

Collection System Operation, Maintenance, & Restoration

Equipment Operation, Evaluation, & Maintenance

64. **d.** The process is more difficult than that for the continuous rod machine

Ref.: page 344, Section 5.3.1.6 Recovering Broken Rods, O & M of Wastewater Collection Systems, Volume 1, 8th edition

Wastewater Collections, Introductory (Grades 1 and 2)

Equipment Operation, Evaluation, & Maintenance

65. **a.** The coupling tool stays connected when a section of the continuous rod machine breaks

Ref.: page 345, Section 5.3.1.6 Recovering Broken Rods, O & M of Wastewater Collection Systems, Volume 1, 8th edition

Wastewater Collections, Introductory (Grades 1 and 2)

Collection System Operation, Maintenance, & Restoration

Equipment Operation, Evaluation, & Maintenance

66. **d.** Are not used for emergencies at night and on weekends

Ref.: page 347, Section 5.3.2 Hand Rods, O & M of Wastewater Collection Systems, Volume 1, 8th edition

Wastewater Collections, Introductory (Grades 1 and 2)

Collection System Operation, Maintenance, & Restoration

Equipment Operation, Evaluation, & Maintenance

67. **b.** Push in tool

Ref.: page 350, Figure 5.14, O & M of Wastewater Collection Systems, Volume 1, 8th edition

Wastewater Collections, Introductory (Grades 1 and 2)

Equipment Operation, Evaluation, & Maintenance

68. **b.** It is easier to rod into an overflowing upstream manhole

Ref.: page 351, numbered items, O & M of Wastewater Collection Systems, Volume 1, 8th edition

Wastewater Collections, Introductory (Grades 1 and 2)

Collection System Operation, Maintenance, & Restoration

Equipment Operation, Evaluation, & Maintenance

69. **a.** The proper tool will be dependent on the sewer's condition

Ref.: page 352, Section 5.3.2.3 Operation, 3rd paragraph, O & M of Wastewater Collection Systems, Volume 1, 8th edition

Wastewater Collections, Introductory (Grades 1 and 2)

Collection System Operation, Maintenance, & Restoration

Equipment Operation, Evaluation, & Maintenance

70. **d.** In a pressurized well-maintained force main

Ref.: page 354, Section 5.3.3 Power Bucket Machines, 1st paragraph, O & M of Wastewater Collection Systems, Volume 1, 8th edition

Wastewater Collections, Introductory (Grades 1 and 2)

Collection System Monitoring, Evaluation, & Adjustment

Collection System Operation, Maintenance, & Restoration

71. **c.** Not a better option than high velocity cleaners in failing mains

    Ref.: page 354, Section 5.3.3 Power Bucket Machines, 4th paragraph, O & M of Wastewater Collection Systems, Volume 1, 8th edition

    Wastewater Collections, Introductory (Grades 1 and 2)

    Collection System Operation, Maintenance, & Restoration

    Equipment Operation, Evaluation, & Maintenance

72. **d.** In smaller-diameter pipes

    Ref.: page 354, Section 5.3.3 Power Bucket Machines, 5th paragraph, O & M of Wastewater Collection Systems, Volume 1, 8th edition

    Wastewater Collections, Introductory (Grades 1 and 2)

    Collection System Operation, Maintenance, & Restoration

    Equipment Operation, Evaluation, & Maintenance

73. **d.** Slack jack

    Ref.: page 357, Figure 5.18, O & M of Wastewater Collection Systems, Volume 1, 8th edition

    Wastewater Collections, Introductory (Grades 1 and 2)

    Equipment Operation, Evaluation, & Maintenance

74. **a.** To prolong equipment life and maintain efficiency

    Ref.: page 365, Section 5.4 Cleaning Equipment and Maintenance, 3rd paragraph, O & M of Wastewater Collection Systems, Volume 1, 8th edition

    Wastewater Collections, Introductory (Grades 1 and 2)

    Equipment Operation, Evaluation, & Maintenance

75. **b.** Partially full hold tank at the end of a run

    Ref.: page 366, Section 5.4.1 High Velocity Cleaners (HVC) Maintenance, O & M of Wastewater Collection Systems, Volume 1, 8th edition

    Wastewater Collections, Introductory (Grades 1 and 2)

    Equipment Operation, Evaluation, & Maintenance

76. **c.** The tension of the drive head chain must be loose

    Ref.: page 367, Section 5.4.2.2 Sectional Rodders, O & M of Wastewater Collection Systems, Volume 1, 8th edition

    Wastewater Collections, Introductory (Grades 1 and 2)

    Equipment Operation, Evaluation, & Maintenance

77. **a.** Check the drive rollers and rods for signs of slippage

Ref.: page 368, Section 5.4.2.3 Continuous Rodder, O & M of Wastewater Collection Systems, Volume 1, 8th edition

Wastewater Collections, Introductory (Grades 1 and 2)

Equipment Operation, Evaluation, & Maintenance

78. **c.** Look for loose rivets and badly worn bails

Ref.: page 369, Section 5.4.3.2 Procedures, O & M of Wastewater Collection Systems, Volume 1, 8th edition

Wastewater Collections, Introductory (Grades 1 and 2)

Equipment Operation, Evaluation, & Maintenance

79. **d.** Leaking and poor joints

Ref.: page 370, Section 5.5 Chemical Control Measures, O & M of Wastewater Collection Systems, Volume 1, 8th edition

Wastewater Collections, Introductory (Grades 1 and 2)

Collection System Monitoring, Evaluation, & Adjustment

Collection System Operation, Maintenance, & Restoration

80. **a.** Most herbicides, insecticides, fungicides, and rodenticides require registration with the USEPA

Ref.: page 371, Section 5.5.1 Precautions for Selecting and using Chemicals, O & M of Wastewater Collection Systems, Volume 1, 8th edition

Wastewater Collections, Introductory (Grades 1 and 2)

Security, Safety, & Administrative Procedures

Collection System Operation, Maintenance, & Restoration

81. **b.** Cutting or pruning sewer roots prevents thicker, faster, and stronger regrowth

Ref.: page 373, Section 5.5.4 Root Control, O & M of Wastewater Collection Systems, Volume 1, 8th edition

Wastewater Collections, Introductory (Grades 1 and 2)

Collection System Monitoring, Evaluation, & Adjustment

Collection System Operation, Maintenance, & Restoration

82. **d.** All of the above

Ref.: page 374, Section 5.5.4.1 Using Chemicals, 1st paragraph, O & M of Wastewater Collection Systems, Volume 1, 8th edition

Wastewater Collections, Introductory (Grades 1 and 2)

Security, Safety, & Administrative Procedures

Collection System Monitoring, Evaluation, & Adjustment

Collection System Operation, Maintenance, & Restoration

83. **a.** destroys existing roots and prevents regrowth

Ref.: page 375, Section 5.5.4.2 Applying Chemicals, 2nd paragraph, O & M of Wastewater Collection Systems, Volume 1, 8th edition

Wastewater Collections, Introductory (Grades 1 and 2)

Collection System Operation, Maintenance, & Restoration

84. **b.** Spraying a layer or filling a section with foam

Ref.: page 376, 2nd-4th paragraphs, O & M of Wastewater Collection Systems, Volume 1, 8th edition

Wastewater Collections, Introductory (Grades 1 and 2)

Collection System Operation, Maintenance, & Restoration

85. **b.** The method is commonly used in sewer mains ranging from 4-6 in. in diameter

Ref.: page 378, Flooding, O & M of Wastewater Collection Systems, Volume 1, 8th edition

Wastewater Collections, Introductory (Grades 1 and 2)

Collection System Operation, Maintenance, & Restoration

86. **d.** All of the above

Ref.: page 380, Section 5.5.5 Grease Control, 2nd paragraph, O & M of Wastewater Collection Systems, Volume 1, 8th edition

Wastewater Collections, Introductory (Grades 1 and 2)

Collection System Monitoring, Evaluation, & Adjustment

Collection System Operation, Maintenance, & Restoration

87. **d.** Pumping the liquid out of the trap or interceptor removes problems

Ref.: page 380, Section 5.5.5 Grease Control, 2nd-4th paragraphs, O & M of Wastewater Collection Systems, Volume 1, 8th edition

Wastewater Collections, Introductory (Grades 1 and 2)

Collection System Monitoring, Evaluation, & Adjustment

Collection System Operation, Maintenance, & Restoration

88. **c.** Identify, prevent, and treat

Ref.: page 382, Section 5.5.6 Rodent and Insect Control, O & M of Wastewater Collection Systems, Volume 1, 8th edition

Wastewater Collections, Introductory (Grades 1 and 2)

Collection System Monitoring, Evaluation, & Adjustment

Collection System Operation, Maintenance, & Restoration

Lift Station Operation & Maintenance

89. **a.** Performing the effort as a separate process from other maintenance activities

Ref.: page 382, Section 5.5.6 Rodent and Insect Control, last paragraph, O & M of Wastewater Collection Systems, Volume 1, 8th edition

Wastewater Collections, Introductory (Grades 1 and 2)

Collection System Monitoring, Evaluation, & Adjustment

Collection System Operation, Maintenance, & Restoration

Lift Station Operation & Maintenance

90. **d.** All of the above

Ref.: page 383, Section 5.6 Odor Control Measures, 2nd paragraph, O & M of Wastewater Collection Systems, Volume 1, 8th edition

Wastewater Collections, Introductory (Grades 1 and 2)

Security, Safety, & Administrative Procedures

Collection System Monitoring, Evaluation, & Adjustment

Collection System Operation, Maintenance, & Restoration

Equipment Operation, Evaluation, & Maintenance

Lift Station Operation & Maintenance

91. **a.** High-velocity flows

Ref.: page 383, Section 5.6 Odor Control Measures, last paragraph, O & M of Wastewater Collection Systems, Volume 1, 8th edition

Wastewater Collections, Introductory (Grades 1 and 2)

Collection System Monitoring, Evaluation, & Adjustment

Collection System Operation, Maintenance, & Restoration

92. **c.** Infrequent monitoring of the sewer H2S concentration

Ref.: page 385, Section 5.6.2 Controlling Odors, O & M of Wastewater Collection Systems, Volume 1, 8th edition

Wastewater Collections, Introductory (Grades 1 and 2)

Collection System Monitoring, Evaluation, & Adjustment

Collection System Operation, Maintenance, & Restoration

93. **b.** Manholes should not be sealed in areas where odors continuously vent

Ref.: pages 386-387, O & M of Wastewater Collection Systems, Volume 1, 8th edition

Wastewater Collections, Introductory (Grades 1 and 2)

Collection System Monitoring, Evaluation, & Adjustment

Collection System Operation, Maintenance, & Restoration

94. **d.** b and c

Ref.: page 390, Carbon Adsorption, O & M of Wastewater Collection Systems, Volume 1, 8th edition

Wastewater Collections, Introductory (Grades 1 and 2)

Security, Safety, & Administrative Procedures

Collection System Monitoring, Evaluation, & Adjustment

Collection System Operation, Maintenance, & Restoration

95. **b.** The wastewater arriving negatively impacts the treatment and generates odors

Ref.: page 402, Section 5.8 Effects of Wastewater Degradation on Wastewater Treatment, O & M of Wastewater Collection Systems, Volume 1, 8th edition

Wastewater Collections, Introductory (Grades 1 and 2)

Collection System Monitoring, Evaluation, & Adjustment

Collection System Operation, Maintenance, & Restoration

Lift Station Operation & Maintenance

96. **d.** All of the above

Ref.: page 412, Section 6.1 Underground Work, O & M of Wastewater Collection Systems, Volume 1, 8th edition

Wastewater Collections, Introductory (Grades 1 and 2)

Security, Safety, & Administrative Procedures

Collection System Operation, Maintenance, & Restoration

Lift Station Operation & Maintenance

97. **c.** A wet tap

Ref.: page 413, Improperly Connected or Defective Taps, O & M of Wastewater Collection Systems, Volume 1, 8th edition

Wastewater Collections, Introductory (Grades 1 and 2)

Collection System Operation, Maintenance, & Restoration

98. **b.** Most utilities do not have specific requirements regarding bedding

Ref.: page 430, Section 6.2.6 Sewer Pipe Bedding and Initial Backfill, O & M of Wastewater Collection Systems, Volume 1, 8th edition

Wastewater Collections, Introductory (Grades 1 and 2)

Collection System Operation, Maintenance, & Restoration

99. **c.** The bell end of the pipe should be facing downstream

Ref.: page 432, Section 6.2.7.2 Placing Pipe Segments, O & M of Wastewater Collection Systems, Volume 1, 8th edition

Wastewater Collections, Advanced (Grades 3 and 4)

Wastewater Collections, Introductory (Grades 1 and 2)

Collection System Operation, Maintenance, & Restoration

100. **d.** All of the above

Ref.: page 434, Section 6.2.8 Sewer Trench Backfill and Compaction, O & M of Wastewater Collection Systems, Volume 1, 8th edition

Wastewater Collections, Introductory (Grades 1 and 2)

Security, Safety, & Administrative Procedures

Collection System Operation, Maintenance, & Restoration

101. **b.** Non-native granular material

Ref.: page 434, Section 6.2.8.1 Backfill Material, O & M of Wastewater Collection Systems, Volume 1, 8th edition

Wastewater Collections, Introductory (Grades 1 and 2)

Collection System Operation, Maintenance, & Restoration

102. **b.** The use of control density fill material

Ref.: page 437, Section 6.2.9 Surface Restoration, O & M of Wastewater Collection Systems, Volume 1, 8th edition

Wastewater Collections, Introductory (Grades 1 and 2)

Collection System Operation, Maintenance, & Restoration

103. **c.** Used instead of rehabilitation or internal inspections

Ref.: page 437, Section 6.2.10 Bypass Piping, O & M of Wastewater Collection Systems, Volume 1, 8th edition

Wastewater Collections, Advanced (Grades 3 and 4)

Wastewater Collections, Introductory (Grades 1 and 2)

Collection System Operation, Maintenance, & Restoration

104. **d.** All of the above

Ref.: page 439, Section 6.3 Sewer Main Line Repairs, O & M of Wastewater Collection Systems, Volume 1, 8th edition

Wastewater Collections, Introductory (Grades 1 and 2)

Collection System Operation, Maintenance, & Restoration

105. **b.** Making sure the frame and cover are secure

Ref.: page 445, Section 6.4 Manhole Repairs, O & M of Wastewater Collection Systems, Volume 1, 8th edition

Wastewater Collections, Introductory (Grades 1 and 2)

Collection System Operation, Maintenance, & Restoration

106. **b.** A wooden check dam

Ref.: page 455, Figure 6.16, O & M of Wastewater Collection Systems, Volume 1, 8th edition

Wastewater Collections, Introductory (Grades 1 and 2)

Collection System Operation, Maintenance, & Restoration

Equipment Operation, Evaluation, & Maintenance

107. **c.** Vent holes are plugged to increase the amount of flow into the system

Ref.: page 456, Section 6.4.5 Preventing Inflow from Entering Through Manhole Covers, 1st and 2nd paragraph, O & M of Wastewater Collection Systems, Volume 1, 8th edition

Wastewater Collections, Introductory (Grades 1 and 2)

Collection System Operation, Maintenance, & Restoration

Equipment Operation, Evaluation, & Maintenance

108. **d.** There is no cleanout in place

Ref.: page 457, Section 6.5 Repairing Sewer Lines, O & M of Wastewater Collection Systems, Volume 1, 8th edition

Wastewater Collections, Introductory (Grades 1 and 2)

Collection System Operation, Maintenance, & Restoration

109. **b.** Insitu flow

Ref.: page 464, Section 6.6 Trenchless Repairs, O & M of Wastewater Collection Systems, Volume 1, 8th edition

Wastewater Collections, Introductory (Grades 1 and 2)

Collection System Operation, Maintenance, & Restoration

110. **b.** Trunk section and rigid fixed rings

Ref.: page 467, Figure 6.24, O & M of Wastewater Collection Systems, Volume 1, 8th edition

Wastewater Collections, Introductory (Grades 1 and 2)

Collection System Operation, Maintenance, & Restoration

Equipment Operation, Evaluation, & Maintenance

111. **c.** High-pressure air is used to test sewers

Ref.: page 474, Section 6.8.5 Air Testing Sewers, O & M of Wastewater Collection Systems, Volume 1, 8th edition

Wastewater Collections, Advanced (Grades 3 and 4)

Collection System Monitoring, Evaluation, & Adjustment

112. **a.** Not recommended to be used for rigid, nonflexible pipe materials

Ref.: page 486, Section 6.8.7 Mandrel Testing Sewers, O & M of Wastewater Collection Systems, Volume 1, 8th edition

Wastewater Collections, Advanced (Grades 3 and 4)

Collection System Monitoring, Evaluation, & Adjustment

Collection System Operation, Maintenance, & Restoration

113. **c.** Lift stations represent a major capital expenditure for any agency or community

Ref.: page 11, Section 8.0 Purpose of Lift Stations, 1st and 2nd paragraph, Operation and Maintenance of Wastewater Collection Systems, Volume II, Seventh edition

Wastewater Collections, Introductory (Grades 1 and 2)

Lift Station Operation & Maintenance

114. **a.** preventing catastrophic failures affecting the community

Ref.: page 11, Section 8.0 Purpose of Lift Stations, 4th paragraph, Operation and Maintenance of Wastewater Collection Systems, Volume II, Seventh edition

Wastewater Collections, Introductory (Grades 1 and 2)

Lift Station Operation & Maintenance

115. **d.** groundwater table is deep and restricts installing deep mains

Ref.: page 11, Section 8.0 0 Location, Operation and Maintenance of Wastewater Collection Systems, Volume II, Seventh edition

Wastewater Collections, Advanced (Grades 3 and 4)

Wastewater Collections, Introductory (Grades 1 and 2)

Lift Station Operation & Maintenance

116. **a.** They are designed to move wastewater with the maximum energy consumption

Ref.: page 12, 1st and 2nd paragraph, Operation and Maintenance of Wastewater Collection Systems, Volume II, Seventh edition

Wastewater Collections, Introductory (Grades 1 and 2)

Lift Station Operation & Maintenance

117. **b.** a manhole equipped with a dry-suction pump

Ref.: page 12, 8.01 Types of Lift Stations, Operation and Maintenance of Wastewater Collection Systems, Volume II, Seventh edition

Wastewater Collections, Advanced (Grades 3 and 4)

Wastewater Collections, Introductory (Grades 1 and 2)

Lift Station Operation & Maintenance

118. **c.** Exposed-electrode controllers

Ref.: page 14, 1st paragraph, Operation and Maintenance of Wastewater Collection Systems, Volume II, Seventh edition

Wastewater Collections, Advanced (Grades 3 and 4)

Wastewater Collections, Introductory (Grades 1 and 2)

Collection System Monitoring, Evaluation, & Adjustment

Lift Station Operation & Maintenance

119. **d.** They are a common type of lift station used in collection systems

Ref.: page 14, Section 8.012 Dry Well Station, 1st paragraph, Operation and Maintenance of Wastewater Collection Systems, Volume II, Seventh edition

Wastewater Collections, Introductory (Grades 1 and 2)

Collection System Operation, Maintenance, & Restoration

Lift Station Operation & Maintenance

120. **d.** The addition of air relief and air vacuum valves

Ref.: page 22, 2nd paragraph, Operation and Maintenance of Wastewater Collection Systems, Volume II, Seventh edition

Wastewater Collections, Advanced (Grades 3 and 4)

Wastewater Collections, Introductory (Grades 1 and 2)

Lift Station Operation & Maintenance

121. **d.** all of the above

Ref.: page 29, Section 8.102 Operating levels, 1st paragraph, Operation and Maintenance of Wastewater Collection Systems, Volume II, Seventh edition

Wastewater Collections, Advanced (Grades 3 and 4)

Wastewater Collections, Introductory (Grades 1 and 2)

Lift Station Operation & Maintenance

122. **b.** are used in upgraded or retrofitted supervisory control systems

Ref.: page 38, 1st paragraph, Operation and Maintenance of Wastewater Collection Systems, Volume II, Seventh edition

Wastewater Collections, Advanced (Grades 3 and 4)

Collection System Monitoring, Evaluation, & Adjustment

123. **d.** Direct current pump motors are the primary ones used

Ref.: page 40, Section 8.140 Types of Motors, Operation and Maintenance of Wastewater Collection Systems, Volume II, Seventh edition

Wastewater Collections, Introductory (Grades 1 and 2)

Equipment Operation, Evaluation, & Maintenance

Lift Station Operation & Maintenance

124. **c.** Impeller, volute, wear rings, and packing rings

Ref.: page 63, Figure 8.40, Typical Non-Clog, Vertical, Raw Wastewater Centrifugal Pump, Operation and Maintenance of Wastewater Collection Systems, Volume II, Seventh edition

Wastewater Collections, Introductory (Grades 1 and 2)

Collection System Operation, Maintenance, & Restoration

Equipment Operation, Evaluation, & Maintenance

125. **b.** Create more wear on the pump shaft or shaft sleeve

Ref.: page 66, Advantages of Mechanical Seals, Operation and Maintenance of Wastewater Collection Systems, Volume II, Seventh edition

Wastewater Collections, Introductory (Grades 1 and 2)

Equipment Operation, Evaluation, & Maintenance

Lift Station Operation & Maintenance

126. **a.** ampere, bearings, insulation, thermistor, and windings

Ref.: page 74, Motor and Control Terms, Operation and Maintenance of Wastewater Collection Systems, Volume II, Seventh edition

Wastewater Collections, Advanced (Grades 3 and 4)

Wastewater Collections, Introductory (Grades 1 and 2)

Collection System Operation, Maintenance, & Restoration

Equipment Operation, Evaluation, & Maintenance

127. **c.** non-rising stem and outside screw and yoke

Ref.: page 75, Section 8.170 Gate Valves, Operation and Maintenance of Wastewater Collection Systems, Volume II, Seventh edition

Wastewater Collections, Introductory (Grades 1 and 2)

Equipment Operation, Evaluation, & Maintenance

128. **b.** adequate drainage

Ref.: page 84, 3rd paragraph, Operation and Maintenance of Wastewater Collection Systems, Volume II, Seventh edition

Wastewater Collections, Advanced (Grades 3 and 4)

Wastewater Collections, Introductory (Grades 1 and 2)

Security, Safety, & Administrative Procedures

Collection System Monitoring, Evaluation, & Adjustment

Lift Station Operation & Maintenance

129. **c.** The potential damage to the station and the surrounding area

Ref.: page 94, Section 8.31 Frequency of Visits to Lift Stations, Operation and Maintenance of Wastewater Collection Systems, Volume II, Seventh edition

Wastewater Collections, Advanced (Grades 3 and 4)

Wastewater Collections, Introductory (Grades 1 and 2)

Lift Station Operation & Maintenance

130. **a.** pumps on line and controlling wet well levels

Ref.: page 95, item 1, top of left side of page, Operation and Maintenance of Wastewater Collection Systems, Volume II, Seventh edition

Wastewater Collections, Advanced (Grades 3 and 4)

Wastewater Collections, Introductory (Grades 1 and 2)

Lift Station Operation & Maintenance

131. **b.** Power, control systems, pumping systems, and structures

Ref.: page 107, bottom left side of page, Operation and Maintenance of Wastewater Collection Systems, Volume II, Seventh edition

Wastewater Collections, Advanced (Grades 3 and 4)

Wastewater Collections, Introductory (Grades 1 and 2)

Lift Station Operation & Maintenance

132. **c.** installing ventilation equipment at the force main outlet

Ref.: page 108, Section 8.33 Troubleshooting Force Main Odors, Operation and Maintenance of Wastewater Collection Systems, Volume II, Seventh edition

Wastewater Collections, Advanced (Grades 3 and 4)

Wastewater Collections, Introductory (Grades 1 and 2)

Security, Safety, & Administrative Procedures

Collection System Monitoring, Evaluation, & Adjustment

133. **a.** Quad phase

Ref.: page 155, Alternating Current, Operation and Maintenance of Wastewater Collection Systems, Volume II, Seventh edition

Wastewater Collections, Introductory (Grades 1 and 2)

Collection System Operation, Maintenance, & Restoration

Equipment Operation, Evaluation, & Maintenance

Lift Station Operation & Maintenance

134. **d.** Proper grounding considers equipment and system components only

Ref.: page 165, Grounding, Operation and Maintenance of Wastewater Collection Systems, Volume II, Seventh edition

Wastewater Collections, Advanced (Grades 3 and 4)

Wastewater Collections, Introductory (Grades 1 and 2)

Security, Safety, & Administrative Procedures

Collection System Operation, Maintenance, & Restoration

Equipment Operation, Evaluation, & Maintenance

Lift Station Operation & Maintenance

135. **b.** Induction motors

Ref.: page 171, Section 9.231 Causes of Failures, #2, Operation and Maintenance of Wastewater Collection Systems, Volume II, Seventh edition

Wastewater Collections, Introductory (Grades 1 and 2)

Equipment Operation, Evaluation, & Maintenance

136. **a.** shaft spacing misalignment

Ref.: page 179, Figure 9.25 Types of Shaft Misalignment and Float, Operation and Maintenance of Wastewater Collection Systems, Volume II, Seventh edition

Wastewater Collections, Introductory (Grades 1 and 2)

Equipment Operation, Evaluation, & Maintenance

137. **c.** cavitation is a result of unusually high pressures in the pump

Ref.: page 198, Cavitation, Operation and Maintenance of Wastewater Collection Systems, Volume II, Seventh edition

Wastewater Collections, Introductory (Grades 1 and 2)

Equipment Operation, Evaluation, & Maintenance

138. **d.** operating at a constant speed within a housing or trough

Ref.: page 202, Section 9.311 Screw Pumps, Operation and Maintenance of Wastewater Collection Systems, Volume II, Seventh edition

Wastewater Collections, Introductory (Grades 1 and 2)

Equipment Operation, Evaluation, & Maintenance

139. **c.** are used to handle limited or low flows with relatively large solids

Ref.: page 203, Section 9.312 Pneumatic Ejectors, Operation and Maintenance of Wastewater Collection Systems, Volume II, Seventh edition

Wastewater Collections, Advanced (Grades 3 and 4)

Collection System Operation, Maintenance, & Restoration

Equipment Operation, Evaluation, & Maintenance

140. **b.** a tankless system

Ref.: page 203, Section 9.312 Pneumatic Ejectors, Operation and Maintenance of Wastewater Collection Systems, Volume II, Seventh edition

Wastewater Collections, Advanced (Grades 3 and 4)

Equipment Operation, Evaluation, & Maintenance

Lift Station Operation & Maintenance

141. **d.** use a fluid cylinder, piston, intake and discharge valves

Ref.: page 205, Section 9.313 Piston Pumps, 1st paragraph, Operation and Maintenance of Wastewater Collection Systems, Volume II, Seventh edition

Wastewater Collections, Introductory (Grades 1 and 2)

Collection System Operation, Maintenance, & Restoration

Equipment Operation, Evaluation, & Maintenance

142. **a.** the smaller stroke allows more water to enter

Ref.: page 205, Section 9.313 Piston Pumps, 2nd paragraph, Operation and Maintenance of Wastewater Collection Systems, Volume II, Seventh edition

Wastewater Collections, Introductory (Grades 1 and 2)

Equipment Operation, Evaluation, & Maintenance

143. **b.** have an extended shaft

Ref.: page 206, Section 9.314 Closed-Coupled Pumps, Operation and Maintenance of Wastewater Collection Systems, Volume II, Seventh edition

Wastewater Collections, Advanced (Grades 3 and 4)

Collection System Operation, Maintenance, & Restoration

Equipment Operation, Evaluation, & Maintenance

144. **a.** its size permits installation in a tight space

Ref.: page 206, Section 9.314 Closed-Coupled Pumps, 2nd and 3rd paragraphs, Operation and Maintenance of Wastewater Collection Systems, Volume II, Seventh edition

Wastewater Collections, Advanced (Grades 3 and 4)

Equipment Operation, Evaluation, & Maintenance

145. **c.** The motor used is waterproof and mechanical seals keep water out

Ref.: page 207, Section 9.315 Submersible Pumps, Operation and Maintenance of Wastewater Collection Systems, Volume II, Seventh edition

Wastewater Collections, Introductory (Grades 1 and 2)

Equipment Operation, Evaluation, & Maintenance

146. **c.** can be hindered by high suction lift

Ref. page 208, Section 9.316 Centrifugal Trash Pumps, Operation and Maintenance of Wastewater Collection Systems, Volume II, Seventh edition

Wastewater Collections, Introductory (Grades 1 and 2)

Equipment Operation, Evaluation, & Maintenance

147. **b.** uses the concentric movement of the rod to alternate from suction to pressure

Ref.: page 210, Section 9.317 Positive Displacement Diaphragm Pumps, Operation and Maintenance of Wastewater Collection Systems, Volume II, Seventh edition

Wastewater Collections, Introductory (Grades 1 and 2)

Equipment Operation, Evaluation, & Maintenance

148. **b.** mixed flow, propeller, radial flow, and turbine

Ref.: page 213, Section 9.40 Impellers, 1st paragraph, Operation and Maintenance of Wastewater Collection Systems, Volume II, Seventh edition

Wastewater Collections, Introductory (Grades 1 and 2)

Equipment Operation, Evaluation, & Maintenance

149. **c.** are closed-type radial or semi-closed for large pumps

Ref.: page 213, Section 9.40 Impellers, 2nd paragraph, Operation and Maintenance of Wastewater Collection Systems, Volume II, Seventh edition

Wastewater Collections, Introductory (Grades 1 and 2)

Collection System Operation, Maintenance, & Restoration

Equipment Operation, Evaluation, & Maintenance

150. **a.** water or wastewater can potentially contaminate the bearings

Ref.: page 214, Section 9.42 Packing, Operation and Maintenance of Wastewater Collection Systems, Volume II, Seventh edition

Wastewater Collections, Introductory (Grades 1 and 2)

Collection System Operation, Maintenance, & Restoration

Equipment Operation, Evaluation, & Maintenance

151. **d.** Minimal leakage at start up

Ref.: page 220, Table 9.7 Causes of Shaft Packing Failure, Operation and Maintenance of Wastewater Collection Systems, Volume II, Seventh edition

Wastewater Collections, Introductory (Grades 1 and 2)

Equipment Operation, Evaluation, & Maintenance

152. **c.** stroke bearings

Ref.: page 223, Section 9.44 Bearings, Operation and Maintenance of Wastewater Collection Systems, Volume II, Seventh edition

Wastewater Collections, Introductory (Grades 1 and 2)

Equipment Operation, Evaluation, & Maintenance

153. **a.** electrical, diesel, gas, or hydraulic pressure

Ref.: page 242, Section 9.450 Purposes of Couplings, Operation and Maintenance of Wastewater Collection Systems, Volume II, Seventh edition

Wastewater Collections, Introductory (Grades 1 and 2)

Equipment Operation, Evaluation, & Maintenance

154. **b.** activating ejectors and pump control systems

Ref.: page 249, Section 9.5 Compressors, Operation and Maintenance of Wastewater Collection Systems, Volume II, Seventh edition

Wastewater Collections, Introductory (Grades 1 and 2)

Collection System Operation, Maintenance, & Restoration

Equipment Operation, Evaluation, & Maintenance

155. **d.** is intended to change the friction between surfaces using the lubricant

Ref.: page 250, Section 9.60 Purpose of Lubrication, 1st and 2nd paragraphs, Operation and Maintenance of Wastewater Collection Systems, Volume II, Seventh edition

Wastewater Collections, Introductory (Grades 1 and 2)

Equipment Operation, Evaluation, & Maintenance

156. **b.** to restore the structural integrity of a failed line

Ref.: page 267, Section 10.0 Historical Background, Operation and Maintenance of Wastewater Collection Systems, Volume II, Seventh edition

Wastewater Collections, Introductory (Grades 1 and 2)

Collection System Monitoring, Evaluation, & Adjustment

Collection System Operation, Maintenance, & Restoration

157. **a.** develop a comprehensive written description of collection system components

Ref.: page 269, Section 10.01 Program Definition, Operation and Maintenance of Wastewater Collection Systems, Volume II, Seventh edition

Wastewater Collections, Introductory (Grades 1 and 2)

Collection System Monitoring, Evaluation, & Adjustment

Collection System Operation, Maintenance, & Restoration

158. **c.** reduce the infiltration/inflow (I/I) problems in existing sewers

Ref.: page 275, Section 10.10 Purpose of Sewer System Evaluations, Operation and Maintenance of Wastewater Collection Systems, Volume II, Seventh edition

Wastewater Collections, Introductory (Grades 1 and 2)

Collection System Monitoring, Evaluation, & Adjustment

159. **d.** have serious negative impacts on the operation of the system

Ref.: page 275, Section 10.11 System Problems, Operation and Maintenance of Wastewater Collection Systems, Volume II, Seventh edition

Wastewater Collections, Introductory (Grades 1 and 2)

Collection System Monitoring, Evaluation, & Adjustment

Collection System Operation, Maintenance, & Restoration

160. **b.** identifying and measuring base flow, infiltration, and inflow

Ref.: page 275, Section 10.12 Hydraulic Aspects, Operation and Maintenance of Wastewater Collection Systems, Volume II, Seventh edition

Wastewater Collections, Advanced (Grades 3 and 4)

Wastewater Collections, Introductory (Grades 1 and 2)

Collection System Monitoring, Evaluation, & Adjustment

Collection System Operation, Maintenance, & Restoration

161. **a.** the potential to contaminate the groundwater in the area

Ref.: page 275, Section 10.12 Hydraulic Aspects, 5th paragraph, Operation and Maintenance of Wastewater Collection Systems, Volume II, Seventh edition

Wastewater Collections, Introductory (Grades 1 and 2)

Collection System Monitoring, Evaluation, & Adjustment

Collection System Operation, Maintenance, & Restoration

162. **c.** house connections, manholes, and sewer mains

Ref.: page 277, bottom right, Operation and Maintenance of Wastewater Collection Systems, Volume II, Seventh edition

Wastewater Collections, Introductory (Grades 1 and 2)

Collection System Monitoring, Evaluation, & Adjustment

Collection System Operation, Maintenance, & Restoration

163. **d.** changes in the land surface over the sewer pipe

Ref.: page 282, Section 10.13 Structural Aspects, Operation and Maintenance of Wastewater Collection Systems, Volume II, Seventh edition

Wastewater Collections, Introductory (Grades 1 and 2)

Collection System Operation, Maintenance, & Restoration

164. **b.** be organized in a format that indicates flow variations by the time of day

Ref.: page 284, Section 10.20 Data Collection, Operation and Maintenance of Wastewater Collection Systems, Volume II, Seventh edition

Wastewater Collections, Advanced (Grades 3 and 4)

Collection System Monitoring, Evaluation, & Adjustment

Collection System Operation, Maintenance, & Restoration

165. **c.** Prepare to replace the worst sections when they fail

Ref.: page 284, Section 10.22 Cost Analysis, 2nd paragraph, Operation and Maintenance of Wastewater Collection Systems, Volume II, Seventh edition

Wastewater Collections, Advanced (Grades 3 and 4)

Collection System Operation, Maintenance, & Restoration

166. **d.** the consequence of failure

Ref.: page 286, Section 10.23 Setting Priorities, 3rd paragraph, Operation and Maintenance of Wastewater Collection Systems, Volume II, Seventh edition

Wastewater Collections, Advanced (Grades 3 and 4)

Wastewater Collections, Introductory (Grades 1 and 2)

Collection System Operation, Maintenance, & Restoration

167. **a.** The selection of pipe materials

Ref.: page 288, Problems Common to All Construction, Operation and Maintenance of Wastewater Collection Systems, Volume II, Seventh edition

Wastewater Collections, Introductory (Grades 1 and 2)

Security, Safety, & Administrative Procedures

Collection System Operation, Maintenance, & Restoration

168. **a.** for sealing leaking pipe joints and circumferential cracks

Ref.: page 290, Section 10.31 Chemical Grouting, Operation and Maintenance of Wastewater Collection Systems, Volume II, Seventh edition

Wastewater Collections, Advanced (Grades 3 and 4)

Wastewater Collections, Introductory (Grades 1 and 2)

Collection System Operation, Maintenance, & Restoration

169. **d.** diamond slotted drill

Ref.: page 294, Figure 10.29, Operation and Maintenance of Wastewater Collection Systems, Volume II, Seventh edition

Wastewater Collections, Introductory (Grades 1 and 2)

Collection System Operation, Maintenance, & Restoration

Equipment Operation, Evaluation, & Maintenance

170. **c.** in downtown areas where there is a high possibility of business disruption

Ref.: page 296, Section 10.320 Pipe Rehabilitation Methods, Operation and Maintenance of Wastewater Collection Systems, Volume II, Seventh edition

Wastewater Collections, Introductory (Grades 1 and 2)

Collection System Operation, Maintenance, & Restoration

171. **b.** a resin-saturated, flexible fabric

Ref.: page 296, Section 10.321 Cured-In-Place Pipe (CIPP), Operation and Maintenance of Wastewater Collection Systems, Volume II, Seventh edition

Wastewater Collections, Advanced (Grades 3 and 4)

Wastewater Collections, Introductory (Grades 1 and 2)

Collection System Operation, Maintenance, & Restoration

172. **c.** identified by a smaller pipe being inserted into a larger pipe

Ref.: page 299, Section 10.322 Sliplining, Operation and Maintenance of Wastewater Collection Systems, Volume II, Seventh edition

Wastewater Collections, Introductory (Grades 1 and 2)

Collection System Operation, Maintenance, & Restoration

173. **c.** reduces the cross-sectional area of the pipeline

Ref.: page 300, Section 10.3220 Continuous High-Density Polyethylene Pipe, Operation and Maintenance of Wastewater Collection Systems, Volume II, Seventh edition

Wastewater Collections, Advanced (Grades 3 and 4)

Wastewater Collections, Introductory (Grades 1 and 2)

Collection System Monitoring, Evaluation, & Adjustment

Collection System Operation, Maintenance, & Restoration

174. **b.** reconditions the sewer system with new, smooth pipe

Ref.: page 313, Section 10.340 Why Line Sewers, Operation and Maintenance of Wastewater Collection Systems, Volume II, Seventh edition

Wastewater Collections, Advanced (Grades 3 and 4)

Wastewater Collections, Introductory (Grades 1 and 2)

Collection System Operation, Maintenance, & Restoration

175. **a.** corrosion or deterioration of the wall and rungs

Ref.: page 320, Section 10.360 Manhole Defects, Operation and Maintenance of Wastewater Collection Systems, Volume II, Seventh edition

Wastewater Collections, Introductory (Grades 1 and 2)

Collection System Monitoring, Evaluation, & Adjustment

Collection System Operation, Maintenance, & Restoration

176. **d.** are used to repair the deteriorated interior of manholes

Ref.: page 325, Section 10.361 Manhole Coatings, Operation and Maintenance of Wastewater Collection Systems, Volume II, Seventh edition

Wastewater Collections, Introductory (Grades 1 and 2)

Collection System Operation, Maintenance, & Restoration

177. **b.** falls, sprains, strains, and being struck by objects

Ref.: page 348, Section 11.2 Types of Accidents and Injuries, Operation and Maintenance of Wastewater Collection Systems, Volume II, Seventh edition

Wastewater Collections, Introductory (Grades 1 and 2)

Security, Safety, & Administrative Procedures

178. **d.** all of the above

Ref.: page 354, Section 11.50 Top Administration, Operation and Maintenance of Wastewater Collection Systems, Volume II, Seventh edition

Wastewater Collections, Advanced (Grades 3 and 4)

Wastewater Collections, Introductory (Grades 1 and 2)

Security, Safety, & Administrative Procedures

179. **a.** safety department staff

Ref.: page 354, Section 11.51 Safety Department Staff, Operation and Maintenance of Wastewater Collection Systems, Volume II, Seventh edition

Wastewater Collections, Advanced (Grades 3 and 4)

Security, Safety, & Administrative Procedures

180. **b.** cover subjects more specific to the work environment

Ref.: page 355, Section 11.6 Monthly Safety Meetings, Operation and Maintenance of Wastewater Collection Systems, Volume II, Seventh edition

Wastewater Collections, Advanced (Grades 3 and 4)

Wastewater Collections, Introductory (Grades 1 and 2)

Security, Safety, & Administrative Procedures

181. **c.** their configuration hinders the worker's access into and out of the space

Ref.: page 358, Section 11.81 Permit-Required Confined Space, Operation and Maintenance of Wastewater Collection Systems, Volume II, Seventh edition

Wastewater Collections, Introductory (Grades 1 and 2)

Security, Safety, & Administrative Procedures

182. **b.** for any employee required to work in a permit-required space

Ref.: page 360, Section 11.814 Training and Education, Operation and Maintenance of Wastewater Collection Systems, Volume II, Seventh edition

Wastewater Collections, Introductory (Grades 1 and 2)

Security, Safety, & Administrative Procedures

183. **c.** being aware of the hazard chemicals used and how to neutralize them

Ref.: page 371, Section 11.10 Homeland Defense, 3rd paragraph, Operation and Maintenance of Wastewater Collection Systems, Volume II, Seventh edition

Wastewater Collections, Introductory (Grades 1 and 2)

Security, Safety, & Administrative Procedures

Collection System Operation, Maintenance, & Restoration

184. **b.** orange

Ref.: page 373, Section 11.13 Threat level categories administered by the US Department of Homeland Security, Operation and Maintenance of Wastewater Collection Systems, Volume II, Seventh edition

Wastewater Collections, Advanced (Grades 3 and 4)

Wastewater Collections, Introductory (Grades 1 and 2)

Security, Safety, & Administrative Procedures

Collection System Operation, Maintenance, & Restoration

185. **c.** the safety committee should review the employee's information and the supervisor's report

Ref.: page 384, Hazard Recognition, Operation and Maintenance of Wastewater Collection Systems, Volume II, Seventh edition

Wastewater Collections, Introductory (Grades 1 and 2)

Security, Safety, & Administrative Procedures

186. **a.** bring unacceptable behavior to the attention of the employee

Ref.: page 388, Disciplinary Policies, Operation and Maintenance of Wastewater Collection Systems, Volume II, Seventh edition

Wastewater Collections, Introductory (Grades 1 and 2)

Security, Safety, & Administrative Procedures

187. **b.** Subpart E, of Part 1910

Ref.: pages 398-399, Applicable Regulations, Operation and Maintenance of Wastewater Collection Systems, Volume II, Seventh edition

Wastewater Collections, Advanced (Grades 3 and 4)

Wastewater Collections, Introductory (Grades 1 and 2)

Security, Safety, & Administrative Procedures

188. **c.** personal protective equipment

Ref.: pages 398-399, Applicable Regulations, Operation and Maintenance of Wastewater Collection Systems, Volume II, Seventh edition

Wastewater Collections, Advanced (Grades 3 and 4)

Wastewater Collections, Introductory (Grades 1 and 2)

Security, Safety, & Administrative Procedures

189. **d.** ensure quality O & M programs to maximize the success of the organization

Ref.: page 427, Section 12.0 Need for Effective Management

Wastewater Collections, Advanced (Grades 3 and 4)

Wastewater Collections, Introductory (Grades 1 and 2)

Security, Safety, & Administrative Procedures

190. **d.** address the agency's mission and goals

Ref.: page 428, Section 12.2 Operating Plan, Operation and Maintenance of Wastewater Collection Systems, Volume II, Seventh edition

Wastewater Collections, Advanced (Grades 3 and 4)

Security, Safety, & Administrative Procedures

Collection System Operation, Maintenance, & Restoration

191. **c.** properly convey wastewater to treatment facilities with minimal disruptions

Ref.: page 428, Section 12.21 Goals, Operation and Maintenance of Wastewater Collection Systems, Volume II, Seventh edition

Wastewater Collections, Advanced (Grades 3 and 4)

Wastewater Collections, Introductory (Grades 1 and 2)

Security, Safety, & Administrative Procedures

Collection System Operation, Maintenance, & Restoration

192. **c.** be clear and concise enough to be reasonably achieved

Ref.: page 428, Section 12.22 Objectives and Tasks/Procedures, Operation and Maintenance of Wastewater Collection Systems, Volume II, Seventh edition

Wastewater Collections, Introductory (Grades 1 and 2)

Security, Safety, & Administrative Procedures

Collection System Operation, Maintenance, & Restoration

193. **b.** consider what is needed to effective manage the assets of the organization

Ref.: page 431, Section 12.3 Personnel, Operation and Maintenance of Wastewater Collection Systems, Volume II, Seventh edition

Wastewater Collections, Advanced (Grades 3 and 4)

Wastewater Collections, Introductory (Grades 1 and 2)

Security, Safety, & Administrative Procedures

194. **a.** establish standards for the effective selection of personnel

Ref.: page 434, Section 12.31 Employment, Operation and Maintenance of Wastewater Collection Systems, Volume II, Seventh edition

Wastewater Collections, Advanced (Grades 3 and 4)

Security, Safety, & Administrative Procedures

195. **d.** all of the above

Ref.: page 436, Section 12.312 Selection, Operation and Maintenance of Wastewater Collection Systems, Volume II, Seventh edition

Wastewater Collections, Advanced (Grades 3 and 4)

Security, Safety, & Administrative Procedures

196. **b.** to recognize the importance of the knowledge of effective O & M practices

Ref.: page 441, Section 12.332 Certification, Operation and Maintenance of Wastewater Collection Systems, Volume II, Seventh edition

Wastewater Collections, Introductory (Grades 1 and 2)

Security, Safety, & Administrative Procedures

Collection System Operation, Maintenance, & Restoration

Lift Station Operation & Maintenance

197. **a.** the location of utility staff

Ref.: page 445, Section 12.40 Basis for Requirements, Operation and Maintenance of Wastewater Collection Systems, Volume II, Seventh edition

Wastewater Collections, Introductory (Grades 1 and 2)

Equipment Operation, Evaluation, & Maintenance

198. **c.** are straight hydraulic jetting units and combination units

Ref.: page 448, High Velocity Cleaners, Operation and Maintenance of Wastewater Collection Systems, Volume II, Seventh edition

Wastewater Collections, Introductory (Grades 1 and 2)

Equipment Operation, Evaluation, & Maintenance

199. **b.** are selected based on the various needs of the organization

Ref.: page 449, Closed-Circuit Television, Operation and Maintenance of Wastewater Collection Systems, Volume II, Seventh edition

Wastewater Collections, Introductory (Grades 1 and 2)

Collection System Monitoring, Evaluation, & Adjustment

Collection System Operation, Maintenance, & Restoration

Equipment Operation, Evaluation, & Maintenance

200. **a.** is based on when and how each piece of equipment is to be maintained

Ref.: page 455, Section 12.432 Preventive Maintenance and Repair, Operation and Maintenance of Wastewater Collection Systems, Volume II, Seventh edition

Wastewater Collections, Introductory (Grades 1 and 2)

Equipment Operation, Evaluation, & Maintenance

201. **c.** when it ceases to perform its required work

Ref.: page 456, Section 12.433 Replacement, Operation and Maintenance of Wastewater Collection Systems, Volume II, Seventh edition

Wastewater Collections, Introductory (Grades 1 and 2)

Equipment Operation, Evaluation, & Maintenance

202. **d.** should be considered for de-centralization based on the least-cost analysis

Ref.: page 457, Section 12.5 Facilities, Operation and Maintenance of Wastewater Collection Systems, Volume II, Seventh edition

Wastewater Collections, Advanced (Grades 3 and 4)

Security, Safety, & Administrative Procedures

Collection System Operation, Maintenance, & Restoration

203. **b.** provide an accurate record of underground system components

Ref.: page 461, Section 12.6 Mapping, Operation and Maintenance of Wastewater Collection Systems, Volume II, Seventh edition

Wastewater Collections, Introductory (Grades 1 and 2)

Security, Safety, & Administrative Procedures

Collection System Monitoring, Evaluation, & Adjustment

Collection System Operation, Maintenance, & Restoration

204. **c.** Building sewers, manholes, and sewer mains

Ref.: page 461, Section 12.610 Locations, Operation and Maintenance of Wastewater Collection Systems, Volume II, Seventh edition

Wastewater Collections, Introductory (Grades 1 and 2)

Collection System Operation, Maintenance, & Restoration

205. **c.** can control cost to operate and control the collection system

Ref.: page 474, Section 12.7031 Types of Work Reports, Operation and Maintenance of Wastewater Collection Systems, Volume II, Seventh edition

Wastewater Collections, Advanced (Grades 3 and 4)

Security, Safety, & Administrative Procedures

Collection System Operation, Maintenance, & Restoration

206. **c.** allows for information to be rapidly manipulated to be accessible and properly used

Ref.: page 488, Section 12.710 Computerized Maintenance Management Systems (CMMS), Operation and Maintenance of Wastewater Collection Systems, Volume II, Seventh edition

Wastewater Collections, Advanced (Grades 3 and 4)

Security, Safety, & Administrative Procedures

Collection System Operation, Maintenance, & Restoration

207. **c.** several databases that collaborate to effectively manage assets

Ref.: page 488, 1st and 2nd paragraph, right side of page, Operation and Maintenance of Wastewater Collection Systems, Volume II, Seventh edition

Wastewater Collections, Advanced (Grades 3 and 4)

Wastewater Collections, Introductory (Grades 1 and 2)

Security, Safety, & Administrative Procedures

Collection System Monitoring, Evaluation, & Adjustment

Collection System Operation, Maintenance, & Restoration

Equipment Operation, Evaluation, & Maintenance

Lift Station Operation & Maintenance

208. **b.** is a program that combines mapping with detailed information about assets

Ref.: page 502, Section 12.711 Geographic Information System (GIS), Operation and Maintenance of Wastewater Collection Systems, Volume II, Seventh edition

Wastewater Collections, Advanced (Grades 3 and 4)

Security, Safety, & Administrative Procedures

Collection System Monitoring, Evaluation, & Adjustment

Collection System Operation, Maintenance, & Restoration

Equipment Operation, Evaluation, & Maintenance

Lift Station Operation & Maintenance

209. **c.** alarm, control, data acquisition, and response

Ref.: page 503, Section 12.710 SCADA, Operation and Maintenance of Wastewater Collection Systems, Volume II, Seventh edition

Wastewater Collections, Introductory (Grades 1 and 2)

Security, Safety, & Administrative Procedures

Collection System Monitoring, Evaluation, & Adjustment

Collection System Operation, Maintenance, & Restoration

Equipment Operation, Evaluation, & Maintenance

Lift Station Operation & Maintenance

210. **a.** to comply with state and federal regulations

Ref. page 515, Section 12.722 Retention of Records, Operation and Maintenance of Wastewater Collection Systems, Volume II, Seventh edition

Wastewater Collections, Introductory (Grades 1 and 2)

Security, Safety, & Administrative Procedures

Collection System Operation, Maintenance, & Restoration

211. **d.** documents the work performed by field crews and work teams

Ref.: page 517, Section 12.80 Why You Need to Write Reports, Operation and Maintenance of Wastewater Collection Systems, Volume II, Seventh edition

Wastewater Collections, Introductory (Grades 1 and 2)

Collection System Operation, Maintenance, & Restoration

212. **b.** providing a service to the public with courtesy, sincerity, and transparency

Ref.: page 519, Section 12.90 Importance of Public Relations, Operation and Maintenance of Wastewater Collection Systems, Volume II, Seventh edition

Wastewater Collections, Advanced (Grades 3 and 4)

Security, Safety, & Administrative Procedures

Collection System Operation, Maintenance, & Restoration

213. **c.** should identify the function of each element and its relationship to O & M

Ref.: page 537, Section 13 Need for Organization, Operation and Maintenance of Wastewater Collection Systems, Volume II, Seventh edition

Wastewater Collections, Advanced (Grades 3 and 4)

Security, Safety, & Administrative Procedures

214. **b.** to allow decisions to be made at the lowest practical level of the organization

Ref.: page 538, 2nd-5th paragraph, left side of page, Operation and Maintenance of Wastewater Collection Systems, Volume II, Seventh edition

Wastewater Collections, Advanced (Grades 3 and 4)

Security, Safety, & Administrative Procedures

215. **d.** establish work standards as well as ways to measure performance

Ref.: page 538, Section 13.15 Accountability, Operation and Maintenance of Wastewater Collection Systems, Volume II, Seventh edition

Wastewater Collections, Advanced (Grades 3 and 4)

Security, Safety, & Administrative Procedures

216. **a.** requires the manager, or supervisor to plan, organize, and staff to efficiently control activities

Ref.: page 539, Section 13.16 Management, Operation and Maintenance of Wastewater Collection Systems, Volume II, Seventh edition

Wastewater Collections, Advanced (Grades 3 and 4)

Security, Safety, & Administrative Procedures

217. **b.** the scope of activities, program goals, objectives, and geographic locations

Ref.: page 540, Section 13.23 Types of Crews, Operation and Maintenance of Wastewater Collection Systems, Volume II, Seventh edition

Wastewater Collections, Advanced (Grades 3 and 4)

Security, Safety, & Administrative Procedures

Collection System Operation, Maintenance, & Restoration

218. **d.** is demonstrated by their success with removing main/service line blockages and correcting lift station failures

Wastewater Collections, Advanced (Grades 3 and 4)

Wastewater Collections, Introductory (Grades 1 and 2)

Security, Safety, & Administrative Procedures

Collection System Operation, Maintenance, & Restoration

Lift Station Operation & Maintenance

219. **c.** A missing manhole cover

Ref.: page 551, Section 13.300 Responding to Requests for Stoppage Removals, Operation and Maintenance of Wastewater Collection Systems, Volume II, Seventh edition

Wastewater Collections, Advanced (Grades 3 and 4)

Security, Safety, & Administrative Procedures

Collection System Operation, Maintenance, & Restoration

220. **b.** should be addressed by staff following approved agency policies

Ref.: page 553, Section 13.301 Flooding Caused by Stoppages, Operation and Maintenance of Wastewater Collection Systems, Volume II, Seventh edition

Wastewater Collections, Advanced (Grades 3 and 4)

Wastewater Collections, Introductory (Grades 1 and 2)

Security, Safety, & Administrative Procedures

Collection System Operation, Maintenance, & Restoration

221. **a.** tends to reduce the number of stoppages and often restores full flow capacity

Ref.: page 555, Section 13.304 Preventive Maintenance, Operation and Maintenance of Wastewater Collection Systems, Volume II, Seventh edition

Wastewater Collections, Introductory (Grades 1 and 2)

Collection System Monitoring, Evaluation, & Adjustment

Collection System Operation, Maintenance, & Restoration

222. **a.** provides a platform to demonstrate how to efficiently and safely perform assigned tasks

Ref.: page 558, Section 13.307 Training Crews, Operation and Maintenance of Wastewater Collection Systems, Volume II, Seventh edition

Wastewater Collections, Advanced (Grades 3 and 4)

Security, Safety, & Administrative Procedures

Collection System Operation, Maintenance, & Restoration

223. **a.** should be based on priority and the need for an immediate repair

Ref.: page 560, Section 13.310 Repairs, Operation and Maintenance of Wastewater Collection Systems, Volume II, Seventh edition

Wastewater Collections, Introductory (Grades 1 and 2)

Security, Safety, & Administrative Procedures

Collection System Operation, Maintenance, & Restoration

Lift Station Operation & Maintenance

224. **b.** improves the reliability of the asset and results in fewer component/system failures

Ref. page 561, Section 13.32 Lift Station Operation and Maintenance, Operation and Maintenance of Wastewater Collection Systems, Volume II, Seventh edition

Wastewater Collections, Introductory (Grades 1 and 2)

Lift Station Operation & Maintenance

225. **c.** are made to verify compliance with the agency's sewer-use and industrial waste ordinances

Ref.: page 567, Section 13.333 Industrial Waste Inspections (Pretreatment Facility Inspections) Operation and Maintenance of Wastewater Collection Systems, Volume II, Seventh edition

Wastewater Collections, Introductory (Grades 1 and 2)

Security, Safety, & Administrative Procedures

Collection System Monitoring, Evaluation, & Adjustment

226. **d.** are a result of conditions developed that interrupt flow and diminish the design capabilities

Ref.: page 573, Section 13.6 Establishing a Maintenance Program, Operation and Maintenance of Wastewater Collection Systems, Volume II, Seventh edition

Wastewater Collections, Introductory (Grades 1 and 2)

Security, Safety, & Administrative Procedures

Collection System Operation, Maintenance, & Restoration

227. **c.** Performing supervisory system maintenance

Ref.: page 576-577, Level 1 Maintenance, Operation and Maintenance of Wastewater Collection Systems, Volume II, Seventh edition

Wastewater Collections, Advanced (Grades 3 and 4)

Collection System Operation, Maintenance, & Restoration

228. **d.** all of the above

Ref.: page 587, Section 13.7 General Performance Indicators, Operation and Maintenance of Wastewater Collection Systems, Volume II, Seventh edition

Wastewater Collections, Advanced (Grades 3 and 4)

Security, Safety, & Administrative Procedures

Collection System Operation, Maintenance, & Restoration

229. **c.** were intended to minimize SSOs and CSOs

Ref.: page 605, Section 14.00 Birth of CMOM, Operation and Maintenance of Wastewater Collection Systems, Volume II, Seventh edition

Wastewater Collections, Advanced (Grades 3 and 4)

Security, Safety, & Administrative Procedures

Collection System Operation, Maintenance, & Restoration

230. **b.** inadequate flow capacities, pipe corrosion, and unexpected population growth

Ref.: page 606-607, Section 14.11 What Causes Poor Performance and Deterioration?, Operation and Maintenance of Wastewater Collection Systems, Volume II, Seventh edition

Wastewater Collections, Advanced (Grades 3 and 4)

Security, Safety, & Administrative Procedures

Collection System Monitoring, Evaluation, & Adjustment

Collection System Operation, Maintenance, & Restoration

231. **d.** all of the above

Ref.: page 610, Section 14.23 Financial Management, Operation and Maintenance of Wastewater Collection Systems, Volume II, Seventh edition

Wastewater Collections, Advanced (Grades 3 and 4)

Security, Safety, & Administrative Procedures

232. **a.** an operating ratio as well as a coverage ratio above 1.0

Ref.: page 610 and 612, Section 14.230 Financial Stability, Operation and Maintenance of Wastewater Collection Systems, Volume II, Seventh edition

Wastewater Collections, Advanced (Grades 3 and 4)

Security, Safety, & Administrative Procedures

233. **b.** is intended to make financial reporting for state and local governments more comprehensive and easier to complete

Ref.: page 613, Section 14.235 Government Accounting Standards Board Statement 34 (GASB 34), Operation and Maintenance of Wastewater Collection Systems, Volume II, Seventh edition

Wastewater Collections, Advanced (Grades 3 and 4)

Security, Safety, & Administrative Procedures

234. **c.** is to keep the system performing as designed and intended

Ref.: page 620, Section 14.30 Purpose of Operation and Maintenance, Operation and Maintenance of Wastewater Collection Systems, Volume II, Seventh edition

Wastewater Collections, Introductory (Grades 1 and 2)

Security, Safety, & Administrative Procedures

Collection System Monitoring, Evaluation, & Adjustment

Collection System Operation, Maintenance, & Restoration

Equipment Operation, Evaluation, & Maintenance

Lift Station Operation & Maintenance

II. Wastewater Treatment, Answers

Math

1. **b.** 1,440,000 or 1.44 mgd

   Ref.: Chapter 7, Appendix 1, Basic Science Concepts and Applications, 4th Edition

   1,000 gpm × 1,440 min/d × 1/1,000,000 = 1.44 mgd

   Wastewater Treatment, Introductory (Grades 1 and 2)

   Treatment Process Monitoring, Evaluation, & Adjustment

2. **c.** 22.8 mgd

   Ref: Appendix A Conversion Tables: Basic Science Concepts and Applications, 4th Edition

   There are two ways to calculate this answer.
   a) 1 m^3 is $(3.28\ ft)^3$ or 35.29 ft^3. A cubic foot will hold 7.48 gal of water.
   There are 86,400 s/d. Therefore,
   35.29 × 7.48 × 86,400/1,000,000 = 22.8 mgd
   b) 1 m^3 holds 264.17 gal and 264.17 × 86,400 = 22.82 mgd

   Wastewater Treatment, Introductory (Grades 1 and 2)

   Treatment Process Monitoring, Evaluation, & Adjustment

3. **c.** 24 gal/h

   Ref: Pages 440-445, Basic Science Concepts and Applications, 4th Edition

   7.2 mgd × day/24 h = 0.3 mil gal/h
   0.3 mil gal/h × 3.79 L/gal = 1.14 ML/h
   (1.14 ML/h) × (12 mg/L) = 13.63 × 10^6 mg = 13.63 × 10^3 g
   Because the polymer is 15 percent strength, at a density of 1.00 g/mL, the required dose is
   13.63 × 10^3/0.15 (strength of solution) = 90.8 ×10^3 mL/h = 90.8 L/h
   or
   300,000 gal/h × 8.34 lb/gal × 12 parts/1,000,000 parts = 30 lb/h polymer dose/h
   (30 lb/h)/0.15 (strength of solution) = 200 lb of solution/h
   (200 lb/h)/(8.34 lb/gal) = 24 gal/h
   Check: (90.8 L/h)/(3.785 L/gal) = 24 gal/h

   Wastewater Treatment, Advanced (Grades 3 and 4)

   Treatment Process Monitoring, Evaluation, & Adjustment

4. **c.** 13.67 m^3 or 3,614 gal

Ref.: Page 54, Operation of Wastewater Treatment Plants Vol II, 7th Edition

The volume of the tank is 69.0 × 10^4 ft^3. That is 690,000/35.3 ft^3/m^3 = 19,538 m^3
19,538 m^3 × 1,000 kg/m^3 × 1.05 g/cm^3 = 20,515,551 kg total weight
Concentration of MLSS removed is 4,200 − 3,500 = 700 mg/L = 7 × 10^{-4} kg/kg
Aeration basin is 20,515,551 kg × 0.7 × 10^{-4} kg/kg removed = 14,360.88 kg removed
Volume removed = 14,360.88/1,050 = 13,677.0 L = 13.670 m^3

Another way of solving the problem:
Aeration basin is 690,000 ft^3 × 62.4 lb/ft^3 × 1.05 = 4,520,880 lb
Weight removed = 4,200 − 3,500 mg/L = 700 mg/L, and because 1 L weighs 10^6 mg, the amount removed is 700/10^6 or 0.007 lb/lb = 4,520,880 × 0.0007 = 31,646 lb removed
Then: 31,646.2 lb is 31,646.2/(8.34 lb/gal × 1.05) = 3,613.8 gal
3,613.8/7.48 (gal/ft^3) = 0.07 percent of volume under aeration

Wastewater Treatment, Advanced (Grades 3 and 4)

Treatment Process Monitoring, Evaluation, & Adjustment

5. **b.** 223,322 gal (845.4 m^3)

Ref.: Page 54, Operation of Wastewater Treatment Plants Vol II, 7th Edition

3,000,000 gal = 11,356,235 L MLSS under aeration is 4,200 mg/L × 11,356,235 L = 47,696,187,000 mg or 47,696.187 kg = 205,217 lb
Amount removed is 35,500 lb/2.20 = 16,102.53 kg = 16,102.53 × 10^6 mg
Solids concentration is 2 percent or 20,000 mg/L
Fluid volume to be removed = 16,102,530 × 1.05/20,000 = 845,368.6 L = 223,322 gal or 845.4 m^3

Wastewater Treatment, Advanced (Grades 3 and 4)

Treatment Process Monitoring, Evaluation, & Adjustment

6. **b.** MLSS 6,930 mg/L, MLVSS 3,410 mg/L

Ref.: Page 243, Operation of Wastewater Treatment Plants, Vol I, 8th Edition

Dry weight of filter + container = 3.52 + 140 g = 143.52 g
Dry weight of filter + container + filtered sample = 147.3 g
Net weight gain = 154.23 − 147.3 for suspended solids = 6.93 g
Assuming that the filter is volatilized, the net gain from MLVSS is 6.93 − 3.52 = 3.41 g
Net weight gain for volatile solids = 6.93 − 3.52 = 3.41 g
MLSS concentration is 6.93 g/1,000 mL = 6,930 mg/L = 6,930 mg/L is MLSS
MLVSS concentration is 3.41 g/1,000 mL = 3,410 mg/L = 3,410 mg/L is MLVSS

Wastewater Collections, Introductory (Grades 1 and 2)

Treatment Process Monitoring, Evaluation, & Adjustment

Laboratory Analysis

7. **b.** 3.18×10^6 m^3/d, 36.81 m^3/s, 840.1 mgd

Ref: Page 781, Operation of Wastewater Treatment Plants Vol. II, 7th Edition

1 ft^3/s is 7.48 gal/ft^3 × 86,400 s/d = 646,272 gpd
1,300 ft^3/s × 86,400 s/d = 112,320,000 ft^3/d
1,300 ft^3/s = 840,153,600 gpd or 840.1×10^6 gpd = 840.1 mgd
(112,320,000 ft^3/d)/35.315 ft^3/m^3 = 3,179,951 m^3/d or 3.18×10^6 m^3/d
3.180×10^6 m^3/d/86,400 s/d = 36.805 m^3/s = 36.81 m^3/s

Wastewater Treatment, Introductory (Grades 1 and 2)

Treatment Process Monitoring, Evaluation, & Adjustment

8. **c.** 1.01 mgd

Ref: Section 3.3, Operation of Wastewater Treatment Plants, Vol. I, 8th edition.

$Q = 2.08 \times (0.83)^{1.58}$ ft^3/s = 1.55 ft^3/s = 1.55 × 0.6463 mgd/ft^3/s = 1.01 mgd

Wastewater Collections, Advanced (Grades 3 and 4)

Treatment Process Monitoring, Evaluation, & Adjustment

9. **c.** 18,900 ft^3

Ref.: Example 8.5, Operation of Wastewater Treatment Plants, Vol. I, 8th Edition

Volume of cylinder plus volume of cone = total volume
Cylinder = $14 \times \pi \times 38^2/4$ = 1,134.1 ft^2 × 14 = 15,877 ft^3
Cone = area of base × height/3 = 1,134.1 × 8/3 = 3,024 ft^3
Total volume = 3,024 + 15,877 = 18,901 ft^3
The answer can also be expressed as 141.4×10^3 gal

Wastewater Treatment, Advanced (Grades 3 and 4)

Treatment Process Monitoring, Evaluation, & Adjustment

Equipment Evaluation & Maintenance

10. **b.** 12.1 h aeration basin, 3.5 h clarifiers

Ref.: Pages 199-200 and Page 708, Operation of Wastewater Treatment Plants, Vol I, 8th Edition

The flow of 6.5 mgd is 6.5/7.48 ft^3/d or 869,000 ft^3/d = 36,207 ft^3/h
Aeration basin detention time is 435,000/36,207 = 12.1 h without recycle
Clarifiers: volume = $82^2/4 \times \pi \times 12$ ft^3 per clarifier = 63,372 ft^3
Volume for two clarifiers = 126,744 ft^3
Detention time = 126,744 ft^3/36,207 ft^3/h = 3.5 h total for both clarifiers
Total hydraulic detention time in all units = 12.1 + 3.5 = 1.6 h

Wastewater Treatment, Advanced (Grades 3 and 4)

Treatment Process Monitoring, Evaluation, & Adjustment

11. **d.** 120 min and 18.8 gpm

First, calculate the tank volume: $20 \times 82 \times \pi/4$ ft^3 = 1,005 ft^3, or approximately 50.25 ft^3 per foot of tank.
Next, the partial volumes can be calculated.
Volume at half full: 1,005 $ft^3/2$ = 502.5 ft^3
Volume at 80 percent full: 1,005 $ft^3 \times 80\%$ = 804 ft^3
804 ft^3 – 502.5 ft^3 = 301.5 ft^3
When the tank is half full, the water height is 10 ft. To reach 80 percent full, the water level needs to reach 16 ft in height, which is 6 ft more than halfway. At a fill rate of 20 min/ft and 6 ft left to fill: 20 min/ft × 6 ft = 120 min.
Because the tank diameter is known to be 8 ft, the area and volume of 1 ft of tank can be calculated.
The pumping rate is $\pi(D^2/4) \times 7.48$ gal/ft^3/20 min = 18.799 or 18.8 gpm

Wastewater Treatment, Advanced (Grades 3 and 4)

Treatment Process Monitoring, Evaluation, & Adjustment

Equipment Evaluation & Maintenance

12. **c.** 96 percent

Ref.: Pages 497-498, Operation of Wastewater Treatment Plants, Vol. I, 8th Edition

((influent concentration – effluent concentration)/influent concentration) × 100% or
((In – out)/in.) × 100% = ((250 – 10)/250) × 100% = 240/250 × 100% = 96% removal rate

Wastewater Treatment, Introductory (Grades 1 and 2)

Treatment Process Monitoring, Evaluation, & Adjustment

13. **c.** 13.82 min

Ref.: https://www.lmnoeng.com/Tank/TankTime.php
https://www.chegg.com/homework-help/questions-and-answers/questioni-water-draining-cylindrical-tank-small-exit-nozzle-section-1-velocity-expressed-t-q38286580

Time = $0.4155 \times (6^2/0.25^2) \times (\oplus 12 - \oplus 0) = 0.4155 \times (36/0.0625) \times 3.464 = 829.05$ s
T = 829 s = 829/60 = 13.82 min
Note: The answer was derived from the Bernoulli equation

Wastewater Treatment, Advanced (Grades 3 and 4)

Treatment Process Monitoring, Evaluation, & Adjustment

Equipment Operation

14. **a.** Phosphorous

Ref.: Page 505, Operation of Wastewater Treatment Plants, Vol. I, 8th Edition

COD:N:P ratio using P = 2 would give 250 COD, 40 TKN, and 2 P.

Wastewater Treatment, Advanced (Grades 3 and 4)

Treatment Process Monitoring, Evaluation, & Adjustment

Laboratory Analysis

15. **d.** 0.17

Ref.: Page 260, Section 5.2, Operation of Wastewater Treatment Plants, Vol. I, 8th Edition
The F:M ratio is the amount of food fed/active biomass.
Influent feed: 5,000,000 gal × 8.34 lb/gal × 250/10^6 = 10,425 lb
MLSS – 335,000 ft^3 × 7.48 gal/ft^3 × 8.34 lb/gal × 2,950/10^6 = 61,650 lb
F:M = 10,425/61,650 = 0.169 or 0.17

Wastewater Treatment, Advanced (Grades 3 and 4)

Treatment Process Monitoring, Evaluation, & Adjustment

Laboratory Analysis

16. **b.** 30 min

Ref.: Pages 336 and 578-593, Operation of Wastewater Treatment Plants, Vol. I, 8th Edition
The chlorine tank volume is 7,000 ft^3. The average flow rate is 1,000,000 gpd = 133,690 ft^3/d = 5,570 ft^3/h. At peak flows, the peaking factor is 2.5, and the flow is 13,926 ft^3/h. The detention time for the chlorine contact tank is 7,000/13,926 = 0.502 h at peak flow or 30.15 min; therefore, detention time is adequate.

Wastewater Treatment, Advanced (Grades 3 and 4)

Treatment Process Monitoring, Evaluation, & Adjustment

17. **a.** 216 gph normal and 518 gph peak flow

Ref.: Pages 590-593, Operation of Wastewater Treatment Plants, Vol. I, 8th Edition

The chlorine feed is 11 percent NaOCl that weighs 10 lb/gal (density 1.2). The formula weight for NaOCl is (23 + 16 + 35.45) = 74.45 g/mol, and it has 1.06 g of Cl_2 per liter of NaOCl. The solution is 11 percent strength, so 1 g of NaOCl will require 1/0.11 or 9.09 or 9.1 L of NaOCl.
The dosage for 5 mg/L as Cl_2 requires 5 mg/L at the following flows:
The normal wastewater flow is 5,570 ft^3/h × 62.38 lb/ft^3 = 359,951 lb/h or 0.36 mil lb/h. A 5 mg/L concentration (5 ppm) would require 5 × 0.36 = 1.8 lb of chlorine per hour.
1.8 lb of chlorine would require 453.6 × 1.8 = g/h of NaOCl = 816.48 L or 215.7 gal/h of an 11 percent chlorine solution. The answer is 216 gph when rounded.
At peak flows, that would require 517.7 (518 gph) of solution. That is 8.63 gpm, almost 9 gpm of chlorine.

Wastewater Treatment, Advanced (Grades 3 and 4)

Treatment Process Monitoring, Evaluation, & Adjustment

18. **b.** Upper 24.52, Lower 16.90

Ref.: Pages 650-655, Ch. 18, Operation of Wastewater Treatment Plants, Vol. II, 7th Edition

Formula for average = $\sum(\text{flow}^1 + \text{flow}^2 + \ldots \text{flow}_n)/N$
Formula for standard deviation = $[(\sum(\text{flow}_1 - \text{avg})^2 + (\text{flow}_2 - \text{avg})^2 + \ldots (\text{flow}_n - \text{avg})^2)/N]^{0.5}$
The formula above is used to calculate the standard deviation, or the spread of the data. Deviation is a measure of the spread of the data. It is calculated by computing the average, then subtracting the average from each of the individual values. Next, calculate the square of the difference between the individual values and the average, and sum the total of those numbers. The square root of the sum of the squares of the difference between the individual daily flows and the average flow divided by the number of measurements is the standard deviation for the data set.
For the data set in this question, the average is 24.518 and the standard deviation is 3.811.
Assuming a normal distribution, approximately 95 percent of the sample values should fall between the normal value ±2 × standard deviation, or 24.52 + 7.62 and 24.52 − 7.62, so the correct answer is b) 24.52 and 16.90
Calculations checked on Excel using $N = 14$

Wastewater Treatment, Advanced (Grades 3 and 4)

Security, Safety, & Administrative Procedures

Treatment Process Monitoring, Evaluation, & Adjustment

19. **a.** 0.43 psi

Ref.: Page 585. Appendix A, Basic Science Concepts and Applications, 4th Edition

Back pressure on aeration systems is usually measured in pounds per square inch. The pressure differential that results from a 1-ft water level is 0.433 ft.
1 ft^3 of water weighs 62.4 lb, spaced over 144 $in.^2$ is 62.4/144 = 0.43 lb/in^2 or psi

Wastewater Treatment, Introductory (Grades 1 and 2)

Treatment Process Monitoring, Evaluation, & Adjustment

Equipment Operation

20. **c.** 33.9 ft

Ref.: Page 217, Figure H2-7 and Page 585, Basic Science Concepts and Applications, 4th Edition

Atmospheric pressure is 14.696 psia
Divide that by 0.4333 psi/ft = 33.93 ft

Wastewater Treatment, Introductory (Grades 1 and 2)

Equipment Operation

21. **d.** Approximately 26 days

Ref.: Pages 21, 54, 58, 67, 70-72, Operation of Wastewater Treatment Plants, Vol. II, 7th Edition

MLSS under aeration: 500,000 gal × 8.34 lb/gal = 4.17×10^6 lb
Solids removed: 15,000 gal × 8.34 lb/gal × 1/1,000,000 = 0.13×10^6 lb
= $0.13 \times 10^6 \times 4{,}500$ mg/10^6 mg = 562.95 lb/day
Sludge age: $4.17 \times 10^6/562.95 \times 10^6$ = 25.92 days, or approximately 26 days

Wastewater Treatment, Advanced (Grades 3 and 4)

Treatment Process Monitoring, Evaluation, & Adjustment

Laboratory Analysis

22. **c.** 110 gpd

Ref.: Pages 794-795, Operation of Wastewater Treatment Plants, Vol I, 8th Edition

2.5×10^6 gal × 8.34 lb/gal × 4 mg/10^6 mg = 83.4 lb of P/day
Gallons required: 1.32 × 83.4 = 110.08 gal

Wastewater Treatment, Advanced (Grades 3 and 4)

Treatment Process Monitoring, Evaluation, & Adjustment

23. **c.** 28,938 ft^3/d

Sludge sent to the digester is 10,000 gal × 9.1 lb/gal = 91,000 lb/d
Solids in sludge = 6 percent or 91,000 × 0.06 = 5,460 lb/d
VS destroyed = 5,460 × (0.65 − 0.12) = 5,460 × 0.53 = 2,893.8 lb destroyed
Gas produced = 10 ft^3/lb destroyed × 2,893.8 lb destroyed = 28,938 ft^3/d

Wastewater Treatment, Advanced (Grades 3 and 4)

Treatment Process Monitoring, Evaluation, & Adjustment

24. **b.** 15,048 lb

Initial feed is 12,000 gpd at 9.5 lb/gal and 12 percent solids = 13,680 lb
Final weight of sludge is 13,680 lb + 10 percent moisture = 15,048 lb

Wastewater Treatment, Advanced (Grades 3 and 4)

Treatment Process Monitoring, Evaluation, & Adjustment

Laboratory Analysis

25. **c.** 792 gpd/ft^2

Ref.: Pages 200-201, Operation of Wastewater Treatment Plants, Vol. I, 8th Edition

Area of clarifier = $75^2/4 \times \pi$ = 4,417.86 ft^2
Overflow rate = 3,500,000/4,417.86
= 792.24 gpd/ft^2

Wastewater Collections, Introductory (Grades 1 and 2)

Treatment Process Monitoring, Evaluation, & Adjustment

Equipment Operation

26. **b.** 350 ft

Ref.: Pages 200-201, Operation of Wastewater Treatment Plants, Vol. I, 8th Edition

3,500,000 gpd/10,000 gpd/ft = 350 ft
The perimeter of a 150-ft-diameter circle is 3.14 × 150 or π × D = 471.2 ft
471.2 ft × 2 clarifiers = 942.5 ft
This is more than adequate by about 2.5 times.
Even adjusting for the width (1 ft) of the launder ring trough, the diameter of the inside of the launder ring would be 148 ft, giving a circumference of 464.96 ft. Two clarifiers would be 929.91 ft, and the launder ring overflow rate would be 3,763.8 gal/ft

Wastewater Treatment, Advanced (Grades 3 and 4)

Treatment Process Monitoring, Evaluation, & Adjustment

27. **a.** 1.91 mg/L dose = 1.41 mg/L demand

Ref.: Page 592, Operation of Wastewater Treatment Plants, Vol. I, 8th Edition

Chlorine dose is: Dose (mg/L) = (feed)/flow
Conversion factor 1 mg/L = 7.48 lb/(1,000,000) gal
Feed = 25/(7.48 × 1.32) = 25/13.07 = 1.91 mg/L dose
Demand = applied dose − residual = 1.91 − 0.5 = 1.41 mg/L demand

Wastewater Treatment, Advanced (Grades 3 and 4)

Treatment Process Monitoring, Evaluation, & Adjustment

28. **c.** 7.2 mg/L/h

Ref:: Page 262-263, Operation of Wastewater Treatment Plants Vol. I, 8th Edition

Oxygen uptake rate is the amount of oxygen consumed in a given time.
Initial oxygen concentration: 3 mg/L
Final oxygen concentration: 1.5 mg/L at 15 min and zero at 25 min
Oxygen demand = (3 mg/L −1.5mg/L)/(15 min/60 min) = 1.5 mg/L/0.25 h
Oxygen demand at 15 min = 1.5/0.25 = 4 mg/h
Oxygen demand at 25 min = 3 mg/L (25 min/60 min) = 3/(25/60)
= 7.2 mg/h

Wastewater Collections, Introductory (Grades 1 and 2)

Treatment Process Monitoring, Evaluation, & Adjustment

Laboratory Analysis

General

1. **b.** globe valve

   Ref.: Ch. 15, Operation of Wastewater Treatment Plants, Vol. 2, 7th edition

   Wastewater Treatment, Introductory (Grades 1 and 2)

   Equipment Operation

   Equipment Evaluation & Maintenance

2. **a.** good flow control

   Ref.: Ch. 15, Operation of Wastewater Treatment Plants, Vol. 2, 7th edition

   Wastewater Collections, Introductory (Grades 1 and 2)

   Equipment Operation

3. **d.** check valve

   Ref.: Ch. 15, Operation of Wastewater Treatment Plants, Vol. 2, 7th edition

   Wastewater Treatment, Introductory (Grades 1 and 2)

   Equipment Operation

   Equipment Evaluation & Maintenance

4. **a.** ensure that flow goes in one direction only.

   Ref.: Ch. 15, Operation of Wastewater Treatment Plants, Vol. 2, 7th edition

   A check valve ensures that flow goes only in one direction and it prevents backflow.

   Wastewater Treatment, Introductory (Grades 1 and 2)

   Equipment Operation

5. **c.** plug valve

   Ref.: Ch. 15, Operation of Wastewater Treatment Plants, Vol. 2, 7th edition.

   Wastewater Treatment, Introductory (Grades 1 and 2)

   Equipment Operation

   Equipment Evaluation & Maintenance

6. **a.** pressure relief valve

   Ref.: Ch. 15, Operation of Wastewater Treatment Plants, Vol. 2, 7th edition.

   When the pressure increases above a certain level, the spring compresses and relieves the pressure.

   Wastewater Treatment, Introductory (Grades 1 and 2)

   Equipment Operation

   Equipment Evaluation & Maintenance

7. **b.** ball valve

   Ref.: Ch. 15, Operation of Wastewater Treatment Plants, Vol. 2, 7th edition.

   Wastewater Treatment, Introductory (Grades 1 and 2)

   Equipment Operation

   Equipment Evaluation & Maintenance

8. **a.** rising stem gate valve

   Ref.: Ch. 15, Operation of Wastewater Treatment Plants, Vol. 2, 7th edition.

   Wastewater Treatment, Introductory (Grades 1 and 2)

   Equipment Operation

   Equipment Evaluation & Maintenance

9. **a.** One can determine whether the valve is open or closed by inspection.

   If the stem is visible, the valve is open; if not, it is closed.

   Wastewater Collections, Introductory (Grades 1 and 2)

   Equipment Operation

   Equipment Evaluation & Maintenance

10. **b.** The valve seat/face connection is wiped when the valve is opened or closed, and the valve will cut stringy or long solids and provide good closure.

    Wastewater Treatment, Introductory (Grades 1 and 2)

    Equipment Operation

    Equipment Evaluation & Maintenance

11. **a.** The valves can collect grit and fail to seal properly.

    Ref.: Ch. 15, Operation of Wastewater Treatment Plants, Vol. 2, 7th edition.

    Wastewater Treatment, Introductory (Grades 1 and 2)

    Equipment Operation

    Equipment Evaluation & Maintenance

12. **c.** butterfly valve

    Ref.: Ch. 15, Operation of Wastewater Treatment Plants, Vol. 2, 7th edition.

    Wastewater Treatment, Introductory (Grades 1 and 2)

    Equipment Operation

    Equipment Evaluation & Maintenance

13. **c.** Stringy and large solids can accumulate on the valve face.

Ref.: Ch. 15, Operation of Wastewater Treatment Plants, Vol. 2, 7th edition.

Rags and strings and other noxious solids can accumulate on the valve when it is open, making it difficult to close. The valve is often high torque when attempts are made to close it.

Wastewater Treatment, Introductory (Grades 1 and 2)

Equipment Operation

Equipment Evaluation & Maintenance

14. **c.** A pressure surge can destroy any pumps connected to the system.

Ref.: Ch. 8 and page 569 (definition), Basic Science Concepts and Applications, 4th edition.

Wastewater Treatment, Introductory (Grades 1 and 2)

Security, Safety, & Administrative Procedures

Equipment Operation

15. **d.** All of the above

Ref.: Section 13.4, Ch 13, Operation of Wastewater Treatment Plants, Vol. 2, 7th edition.

All of the above should be recorded before a sample is transmitted to the laboratory for compliance evaluation.

Wastewater Treatment, Introductory (Grades 1 and 2)

Security, Safety, & Administrative Procedures

Laboratory Analysis

16. **b.** to cool the sample to 4°C and analyze it within 48 h, storing it in glass or plastic.

Ref.: Chapter 17, NEIPS Handbook for Sampling and Sample Preservation
Pages 500-502, Operation of Wastewater Treatment Plant, Vol. 2, 7th edition.
Pages 825-27, Operation of Wastewater Treatment Plant, Vol. 1, 8th edition.

According to the USEPA, most samples should be cooled to 4°C and transported to the lab within 48 h, unless other preservatives are used. The most common material for containing samples is polyethylene with a polypropylene cap. Source: http://sisbl.uga.edu/epatab1.html.

Wastewater Treatment, Introductory (Grades 1 and 2)

Laboratory Analysis

17. **d.** All of the above are true.

Ref.: Ch. 9.4 and 9.5, Operation of Wastewater Treatment Plants, Vol.1, 8th edition.

Flow proportioning samplers do need a flowmeter connection. There are two distinct theories of sample compositing samples. Time proportional collects samples at fixed intervals and the volume of each aliquot is constant. Flow proportional samplers collect one sample that is a large-scale composite of individual flow measurements.

Wastewater Treatment, Introductory (Grades 1 and 2)

Security, Safety, & Administrative Procedures

Treatment Process Monitoring, Evaluation, & Adjustment

Laboratory Analysis

18. **c.** Set up a monitoring program with split sampling (one set of samples analyzed by the utility and one by the company) to verify the results.

Ref.: Pages 813, Ch. 9.4 and pages 886-890, Operation of Wastewater Treatment Plants, Vol. 1, 8th edition. Page 13, Ch. 5, Pennsylvania Department of Environmental Protection.

The reported accuracies (as per Standard Methods) for TSS is about 15 percent (there is no given standard for TSS), and the variance in test results for BOD_5 are quite variable. The BOD_5 test accuracy is also about 15 percent at 200 mg/L and is one of the most inaccurate tests in the environmental monitoring program. Given this, the variability of the test requires multiple samples for any accurate determination.

Wastewater Treatment, Advanced (Grades 3 and 4)

Security, Safety, & Administrative Procedures

Laboratory Analysis

19. **a.** Thoroughly clean and rinse all tubing in contact with liquid, thoroughly clean the sample container, and replace the tubing if necessary.

Wastewater Treatment, Introductory (Grades 1 and 2)

Equipment Evaluation & Maintenance

Laboratory Analysis

20. **b.** Most sampling methods consistently under collect the true discharges by 1 to 10 percent.

Ref.: Pages 815-828, Operation of Wastewater Treatment Plants, Vol. 1, 8th edition.

Under most circumstances, sampling programs understate the actual concentrations of materials in the flow by between 1 and 10 percent.

Wastewater Treatment, Introductory (Grades 1 and 2)

Treatment Process Monitoring, Evaluation, & Adjustment

Laboratory Analysis

21. **a.** 48 hours at 4°C

Ref.: U. S. Environmental Protection Agency. 1983.
Sample preservation. In Methods for Chemical Analysis of Water and Wastes, EPA-600/4-79-020. U.S.E.P.A., Cincinnati, Ohio, USA.
NEIPS Handbook for Sampling and Sample Preservation, Chapter 17.
Ch. 9.4, Operation of Wastewater Treatment Plants, Vol. 1, 8th edition.

Wastewater Treatment, Introductory (Grades 1 and 2)

Security, Safety, & Administrative Procedures

Treatment Process Monitoring, Evaluation, & Adjustment

Laboratory Analysis

22. **a.** Influent, effluent, and waste activated sludge

Definition of a material balance includes sampling at influent, effluent and waste activated sludge to determine the total phosphorous removals.

Wastewater Treatment, Advanced (Grades 3 and 4)

Treatment Process Monitoring, Evaluation, & Adjustment

Laboratory Analysis

23. **d.** All of the above.

Ref.: Pages 886-90, Operation of Wastewater Treatment Plants, Vol. 1, 8th edition.
Standard Methods for the Examination of Water and Wastewater, 23rd edition.

Wastewater Treatment, Introductory (Grades 1 and 2)

Laboratory Analysis

24. **b.** A liquid is incompressible while a gas is compressible.

Ref.: Ch. 15, Operation of Wastewater Treatment Plants, Vol. 2, 7th edition.

Wastewater Treatment, Introductory (Grades 1 and 2)

Treatment Process Monitoring, Evaluation, & Adjustment

Laboratory Analysis

25. **a.** Solids should be wasted to reduce the mixed liquor suspended solids and oxygen demand, and the diffusers should be checked for plugging.

Ref.: Pages 19, 49, and 263, Operation of Wastewater Treatment Plants, Vol. 2, 7th edition.

The optimal level for dissolved oxygen in the aeration tanks is between 1.7 and 2.0 mg/L. The options available for the operator are limited. Additional blowers may not be available. Higher-than-normal influent loadings may be a factor, but there is not much an operator can do about it. Changing diffusers to increase oxygen transfer takes hours to days.

Wastewater Treatment, Advanced (Grades 3 and 4)

Treatment Process Monitoring, Evaluation, & Adjustment

26. **b.** Removing the grit (sand and gravel) to protect pumps and equipment.

Ref.: Section 3.5, pp 136-137, 165-172, 186-187 and 1044, Operation of Wastewater Treatment Plants, Vol. 1, 8th edition.

Wastewater Treatment, Introductory (Grades 1 and 2)

Equipment Operation

Equipment Evaluation & Maintenance

27. **a.** A storm sewer that also conveys sanitary flows.

Ref.: Pages 11, 1039, Operation of Wastewater Treatment Plants, Vol. 1, 8th edition.

Wastewater Treatment, Introductory (Grades 1 and 2)

Treatment Process Monitoring, Evaluation, & Adjustment

28. **b.** Anoxic is without free oxygen (ORP +50 to –50 mv), whereas anaerobic is without any oxygen (–100 to –225 mv).

Ref.:
Page 1036 definitions, Operation of Wastewater Treatment Plants, Vol. 1, 8th edition.
Page 74, footnote 35, Operation of Wastewater Treatment Plants, Vol. 2, 7th edition.

Wastewater Treatment, Advanced (Grades 3 and 4)

Treatment Process Monitoring, Evaluation, & Adjustment

29. **a.** An anoxic environment has no free oxygen but has bound oxygen such as nitrates and phosphates, whereas an anaerobic environment has no oxygen.

Wastewater Treatment, Advanced (Grades 3 and 4)

Treatment Process Monitoring, Evaluation, & Adjustment

30. **b.** Chemical oxygen demand (COD), NH_3, NO_3^-, total suspended solids (TSS), mixed liquor volatile suspended solids (MLVSS), and oxygen uptake rate (OUR)

Ref.: Pages 18-21, Ch. 5, Operation of Wastewater Treatment Plants, Vol. 1, 8th edition.

Analysis for BOD will only identify what happened to the plant five days ago. NH_3 is an oxygen demanding substance, and MLVSS indicates how much active biomass is under aeration. OUR indicates how effectively the biomass is receiving the aeration.

Wastewater Treatment, Advanced (Grades 3 and 4)

Treatment Process Monitoring, Evaluation, & Adjustment

31. **c.** A device to remove suspended solid material from water by sedimentation.

Ref.: Pages 192-208, 243-245, 444-46, 483-5, Operation of Wastewater Treatment Plants, Vol.1, 8th edition.

Wastewater Treatment, Introductory (Grades 1 and 2)

Equipment Operation

Equipment Evaluation & Maintenance

32. **a.** Remove excess water from sludge and concentrate it by stirring.

Ref.: Ch. 3, Operation of Wastewater Treatment Plants, Vol. 2, 7th edition.

The sludge thickener rakes stir the sludge to help it concentrate and dewater it and prevent bridging.

Wastewater Treatment, Introductory (Grades 1 and 2)

Treatment Process Monitoring, Evaluation, & Adjustment

Equipment Operation

Equipment Evaluation & Maintenance

33. **a.** Source of soluble food, nutrients, and oxygen or an electron acceptor.

The food must be soluble to penetrate the microorganism's cell wall.

Wastewater Treatment, Introductory (Grades 1 and 2)

Treatment Process Monitoring, Evaluation, & Adjustment

34. **b.** The rate of bacterial activity decreases by about half with a 10°C (18°F) change in temperature.

Ref.: Pages 21, 54, 64, 70, and 218, Operation of Wastewater Treatment Plants, Vol. 2, 7th edition.

Wastewater Treatment, Introductory (Grades 1 and 2)

Treatment Process Monitoring, Evaluation, & Adjustment

35. **b.** 6–9

Ref.: Pages 862-864, Operation of Wastewater Treatment Plants, Vol. 2, 7th edition.

Bacterial growth can be adversely affected by pH outside of the range of 6–9.

Wastewater Treatment, Introductory (Grades 1 and 2)

Treatment Process Monitoring, Evaluation, & Adjustment

36. **a.** The presence of alkyl benzene sulfonate (old style detergent) in the water.

Ref.: Pages 44, 299 paragraph 14.233, Operation of Wastewater Treatment Plants, Vol. 2, 7th edition.

The use of alkyl benzene sulfonate has been discontinued in the United States by switching to linear alkyl sulfonates. Similarly, the discharge of most commercial detergents won't cause this condition.

Wastewater Treatment, Introductory (Grades 1 and 2)

Treatment Process Monitoring, Evaluation, & Adjustment

37. **c.** Coats and accumulates inside of pipes, create plugs, and interferes with lift stations operations.

Wastewater Treatment, Introductory (Grades 1 and 2)

Treatment Process Monitoring, Evaluation, & Adjustment

38. **a.** Effective enforcement of a grease and oil collection program.

A simple ban does not provide a good solution. A collection program is necessary. There is a federal limit on the concentration of grease and oil in commercial and industrial discharges.

Wastewater Treatment, Introductory (Grades 1 and 2)

Security, Safety, & Administrative Procedures

Treatment Process Monitoring, Evaluation, & Adjustment

39. **c.** The food industry can discharge high levels of organics, which will overwhelm the aeration capacity of the plant. Hospitals can discharge anti-bacterial compounds, which can kill the bacteria.

Wastewater Treatment, Introductory (Grades 1 and 2)

Security, Safety, & Administrative Procedures

Treatment Process Monitoring, Evaluation, & Adjustment

40. **a.** Toxic wastes can kill off the bacteria in the plant and cause a discharge of contaminated water by lowering or stopping bacterial metabolism.

Ref.: Pages 70-72, 854, 299 paragraph (14.234), Operation of Wastewater Treatment Plants, Vol. 2, 7th edition.

Wastewater Treatment, Introductory (Grades 1 and 2)

Security, Safety, & Administrative Procedures

Treatment Process Monitoring, Evaluation, & Adjustment

41. **a.** A form of biosolids stabilization.

Ref.: Ch. 12, Operation of Wastewater Treatment Plants, Vol. 2, 7th edition.

Wastewater Treatment, Introductory (Grades 1 and 2)

Treatment Process Monitoring, Evaluation, & Adjustment

42. **a.** Very low dissolved oxygen (DO), high pH, temperature less than 10°C, low alkalinity

Ref.: Page 126, 239, Chapter 11, Operation of Wastewater Treatment Plants, Vol. 1, 8th edition. Page 21, Operation of Wastewater Treatment Plants, Vol. 2, 7th edition.

Nitrifiers need temperatures between 10°C and 30°C, a pH between 7 and 8.5, a DO well above 1.0 mg/L, and an amount of alkalinity to support the reaction depending on the level of ammonia and nitrite present.

Wastewater Treatment, Advanced (Grades 3 and 4)

Treatment Process Monitoring, Evaluation, & Adjustment

43. **c.** Arsenic and lead are very toxic even at very low levels.

Ref.:
Interdiscip Toxicol. 2014 Jun; 7(2): 60–72.
Experientia Supplementum. 2012; 101: 133–164.

Chromium, copper, lead, arsenic, and zinc can all have a negative effect on the treatment plant depending on the form of the metal ion and the pH of the system. If either the pH or the concentration are in specific ranges, toxicity can result. Chromium is quite toxic in the +6 valence state, but very much less so in the +3 valence state. Zinc can be toxic at high concentrations.

Wastewater Treatment, Advanced (Grades 3 and 4)

Treatment Process Monitoring, Evaluation, & Adjustment

44. **c.** To capture floatable solids and reduce some of the biodegradable load to the aeration chamber.

Ref.: Ch. 4.1, Operation of Wastewater Treatment Plants, Vol. 1, 8th edition.

Wastewater Treatment, Introductory (Grades 1 and 2)

Treatment Process Monitoring, Evaluation, & Adjustment

45. **b.** It is sent to an anaerobic digester for further treatment.

Ref.: Ch. 4.1& 4.5 and pages 211-212, Operation of Wastewater Treatment Plants, Vol. 1, 8th edition.

Wastewater Treatment, Introductory (Grades 1 and 2)

Treatment Process Monitoring, Evaluation, & Adjustment

46. **b.** To concentrate the settleable materials and move them to the collection point.

Ref.: Ch. 4 and page 197, Operation of Wastewater Treatment Plants, Vol. 1, 8th edition.

Wastewater Treatment, Introductory (Grades 1 and 2)

Equipment Operation

Equipment Evaluation & Maintenance

47. **d.** The flow pattern and settling pattern in a rectangular clarifier are linear.

Both types of clarifiers form sludge blankets. The center feed well in a circular clarifier is designed to prevent short circuiting.

Wastewater Treatment, Introductory (Grades 1 and 2)

Treatment Process Monitoring, Evaluation, & Adjustment

Equipment Operation

48. **b.** Organics 20–33 percent, suspended solids up to 40 percent

Ref.: Section 4.7 and page 297, Operation of Wastewater Treatment Pants, Vol. 1, 8th edition.

Wastewater Treatment, Introductory (Grades 1 and 2)

Treatment Process Monitoring, Evaluation, & Adjustment

Equipment Operation

Equipment Evaluation & Maintenance

49. **a.** A baffle to collect the floatable materials and prevent them from going into the main aeration chamber.

Ref.: Page 197, Ch. 4, Operation of Wastewater Treatment Plants, Vol. 1, 8th edition.

Wastewater Treatment, Introductory (Grades 1 and 2)

Equipment Operation

50. **b.** Ensure even distribution of the flow across the clarifier surface.

Ref.: Page 196, Ch. 4, Operation of Wastewater Treatment Plants, Vol. 1, 8th edition.

Wastewater Treatment, Introductory (Grades 1 and 2)

Equipment Operation

51. **a.** The bar screen traps large solids for removal, whereas the comminutor shreds large particles into much smaller particles.

Ref.: Section 3.4, Operation of Wastewater Treatment Plants, Vol. 1, 8th edition.

Wastewater Treatment, Introductory (Grades 1 and 2)

Equipment Operation

Equipment Evaluation & Maintenance

52. **b.** Some of the excess solids are wasted and sent through a thickening /dewatering/ digestion system for ultimate soil amendment or fertilizer use.

Ref.: Ch. 5.3, Operation of Wastewater Treatment Plants, Vol. 1, 8th edition.

Wastewater Treatment, Introductory (Grades 1 and 2)

Treatment Process Monitoring, Evaluation, & Adjustment

53. **b.** It maintains the mixed liquor suspended solids (MLSS) level in the aeration tank.

Ref.: Ch. 5.3, Operation of Wastewater Treatment Plants, Vol. 1, 8th edition.

Wastewater Collections, Introductory (Grades 1 and 2)

Treatment Process Monitoring, Evaluation, & Adjustment

Equipment Operation

54. **c.** It creates small bubbles to increase the gas transfer rates.

Ref.: Pages 172, & 418 and 419, Operation of Wastewater Treatment Plants, Vol. 1, 8th edition.

Wastewater Treatment, Introductory (Grades 1 and 2)

Treatment Process Monitoring, Evaluation, & Adjustment

Equipment Operation

55. **a.** A receptacle for collecting the floating solids from the surface of a clarifier.

Ref.: Ch. 8.7.6, Operation of Wastewater Treatment Plants, Vol. 1, 8th edition

Wastewater Treatment, Introductory (Grades 1 and 2)

Equipment Operation

56. **b.** To redirect the flow in the clarifier toward the bottom of the clarifier.

Ref.: Page 195, Operation of Wastewater Treatment Plants, Vol. 1, 8th edition.

Wastewater Treatment, Introductory (Grades 1 and 2)

Equipment Operation

57. **b.** Coarse bubble allows higher mixed liquor suspended solids (MLSS) in the aeration tank because of better mixing than fine bubbles.

Ref.: Pages 374-76, and 270-76, Operation of Wastewater Treatment Plants, Vol. 1, 8th edition.

Wastewater Treatment, Introductory (Grades 1 and 2)

Treatment Process Monitoring, Evaluation, & Adjustment

Equipment Operation

58. **b.** FOG can greatly add to the biochemical oxygen demand (BOD)/chemical oxygen demand (COD) load of the plant.

Ref.: Section 5.5 (Abnormal Operations), Operation of Wastewater Treatment Plants, Vol. 1, 8th edition.

Wastewater Treatment, Introductory (Grades 1 and 2)

Treatment Process Monitoring, Evaluation, & Adjustment

59. **a.** Low dissolved oxygen (DO) in the aeration basin.

Ref.: Page 243, Operation of Wastewater Treatment Plants, Vol. 1, 8th edition. Pages 19, 61-64, 71-72, Operation of Wastewater Treatment Plants, Vol. 2, 7th edition.

Wastewater Treatment, Advanced (Grades 3 and 4)

Treatment Process Monitoring, Evaluation, & Adjustment

60. **a.** Add chlorine to the secondary clarifier.

Ref.: Pages 71-72, Operation of Wastewater Treatment Plants, Vol. 2, 7th edition.

Wastewater Treatment, Introductory (Grades 1 and 2)

Treatment Process Monitoring, Evaluation, & Adjustment

61. **a.** It is a measure of the net fluid upward velocity that controls the speed of mixed liquor suspended solids (MLSS) settling.

    Ref.: Page 200, Operation of Wastewater Treatment Plants, Vol. 2, 7th edition.

    Wastewater Treatment, Advanced (Grades 3 and 4)

    Treatment Process Monitoring, Evaluation, & Adjustment

62. **c.** Anything that could raise groundwater levels and that increase infiltration and inflow.

    Wastewater Treatment, Introductory (Grades 1 and 2)

    Treatment Process Monitoring, Evaluation, & Adjustment

63. **c.** It is a measure of the acidity or basicity of the water, and the optimal range is between 9 and 6.

    Ref.: Pages 7, 701, 863, Operation of Wastewater Treatment Plants, Vol. 1, 8th edition.

    Wastewater Treatment, Introductory (Grades 1 and 2)

    Treatment Process Monitoring, Evaluation, & Adjustment

64. **b.** Alkalinity is the product of bicarbonates and weak acid salts, and it is required for anaerobic degradation and denitrification.

    Ref.: Pages 848-850, Operation of Wastewater Treatment Plants, Vol. 1, 8th edition.

    Wastewater Treatment, Introductory (Grades 1 and 2)

    Treatment Process Monitoring, Evaluation, & Adjustment

    Laboratory Analysis

65. **b.** It is a form of extended aeration treatment.

    Ref.: Pages 312-314, Pages 830-845, Operation of Wastewater Treatment Plants, Vol. 1, 8th edition.

    Wastewater Treatment, Introductory (Grades 1 and 2)

    Treatment Process Monitoring, Evaluation, & Adjustment

    Equipment Operation

66. **d.** All of the above.

    Ref.: Pages 332-334, Pages 830-845, Operation of Wastewater Treatment Plants, Vol. 1, 8th edition.

    Wastewater Treatment, Introductory (Grades 1 and 2)

    Treatment Process Monitoring, Evaluation, & Adjustment

67. **b.** Septage is another name for the materials taken from a septic tank or holding tank.

    Ref.: Page 731, Pages 830-845, Operation of Wastewater Treatment Plants, Vol. 1, 8th edition.

    Wastewater Treatment, Introductory (Grades 1 and 2)

    Treatment Process Monitoring, Evaluation, & Adjustment

68. **b.** They are an impoundment that encourage natural processes and algae to promote waste treatment.

Ref.: Page 222, Operation of Wastewater Treatment Plants, Vol. 2, 7th edition.

Wastewater Treatment, Introductory (Grades 1 and 2)

Treatment Process Monitoring, Evaluation, & Adjustment

69. **a.** During daylight algae produces O_2 by photosynthesis and raises the pH. At night algae and microorganisms convert the O_2 and give off CO_2, which lowers the pH.

Ref.: Pages 35-36, 732-735, Operation of Wastewater Treatment Plants, Vol. 1, 8th edition.

Wastewater Treatment, Introductory (Grades 1 and 2)

Treatment Process Monitoring, Evaluation, & Adjustment

70. **d.** Stabilization ponds do not treat wastewater as well as an activated sludge system.

Ref.: Page 742, Operation of Wastewater Treatment Plants, Vol. 1, 8th edition.

Wastewater Treatment, Advanced (Grades 3 and 4)

Treatment Process Monitoring, Evaluation, & Adjustment

71. **d.** All of the above.

Ref.: Pages 741-738, Operation of Wastewater Treatment Plants, Vol. 1, 8th edition.

Wastewater Treatment, Introductory (Grades 1 and 2)

Treatment Process Monitoring, Evaluation, & Adjustment

72. **b.** Newer technology using ultrasonic equipment that prevents algae growth.

Ref.: Pages 731-738, Operation of Wastewater Treatment Plants, Vol. 2, 7th edition.

Ultrasonic control systems for algae growth prevention are relatively new as of 2010. The covering of ponds and dyeing of the waters has also been successful but is expensive. Heavy doses of alum or ferric salts can also work, but they are also very expensive.

Wastewater Treatment, Introductory (Grades 1 and 2)

Treatment Process Monitoring, Evaluation, & Adjustment

73. **b.** A growth of filamentous organisms that causes poor settling in clarifiers.

Ref.: Pages 71, 90, and 116-119, Operation of Wastewater Treatment Plants, Vol. 1, 8th edition.

Filamentous organisms, such as Nocardia and others, will, in sufficient quantity, grow in the aeration tank and will form bridges, which have very low settling rates, and cause high suspended solids in the clarifier effluent.

Wastewater Treatment, Introductory (Grades 1 and 2)

Treatment Process Monitoring, Evaluation, & Adjustment

74. **a.** Chlorination and higher recycle that increase the F:M ratio.

Ref.: Page 71, Operation of Wastewater Treatment Plants, Vol. 2, 7th edition.

Wastewater Treatment, Introductory (Grades 1 and 2)

Treatment Process Monitoring, Evaluation, & Adjustment

75. **b.** It removes grease and oil from the influent.

Ref.: Page 32, Operation of Wastewater Treatment Plants, Vol. 1, 8th edition.

The DAF unit can remove suspended solids and, depending on application, can partially substitute for a primary clarifier.

Wastewater Treatment, Advanced (Grades 3 and 4)

Treatment Process Monitoring, Evaluation, & Adjustment

Equipment Operation

76. **a.** Trickling filter, rotating biological contactors, and membrane bioreactors.

Ref.: Section 6.2, Operation of Wastewater Treatment Plants, Vol. 1, 8th edition.

All others are suspended growth systems.

Wastewater Treatment, Introductory (Grades 1 and 2)

Treatment Process Monitoring, Evaluation, & Adjustment

77. **a.** The processes rely on development of a biofilm attached to a media for stability and support.

Wastewater Treatment, Advanced (Grades 3 and 4)

Treatment Process Monitoring, Evaluation, & Adjustment

78. **a.** Both fine and coarse bubble aerators are more efficient at oxygen transfer than surface aerators.

Wastewater Treatment, Introductory (Grades 1 and 2)

Treatment Process Monitoring, Evaluation, & Adjustment

Equipment Operation

79. **a.** They provide good mixing for large areas.

Wastewater Treatment, Introductory (Grades 1 and 2)

Equipment Operation

80. **a.** The purpose of the anaerobic section is to reduce and remove both nitrogen and phosphorous.

Ref.: Page 90, (Bardenpho Process), Operation of Wastewater Treatment Plants, Vol. 1, 8th edition.

Wastewater Treatment, Advanced (Grades 3 and 4)

Treatment Process Monitoring, Evaluation, & Adjustment

81. **a.** Anaerobic is for phosphorus (P) removal, whereas anoxic is for nitrogen (N) removal.

Ref.: Page 90, Operation of Wastewater Treatment Plants, Vol. 2, 7th Edition.

Wastewater Treatment, Advanced (Grades 3 and 4)

Treatment Process Monitoring, Evaluation, & Adjustment

82. **a.** The second figure is a phostrip process where the phosphate is removed by chemical precipitation.

Ref.: Page 90, Operation of Wastewater Treatment Plants, Vol. 2, 7th edition.

Wastewater Treatment, Advanced (Grades 3 and 4)

Treatment Process Monitoring, Evaluation, & Adjustment

83. **b.** Anaerobic action sets in and starts generating gas bubbles, causing poor performance of the clarifier.

Gas production can occur if the dissolved oxygen in the clarifier drops to zero.

Wastewater Treatment, Introductory (Grades 1 and 2)

Treatment Process Monitoring, Evaluation, & Adjustment

84. **a.** Calcium hydroxide

Chlorine does not generate a sludge and cannot remove phosphorous from the water.

Wastewater Treatment, Advanced (Grades 3 and 4)

Treatment Process Monitoring, Evaluation, & Adjustment

85. **a.** 20:1 and 45:1

Ref.: Treatment Plant Operator, "Optimal Nutrient Ratios for Wastewater Treatment", 2018. Brochure. NCBI ,2014; 1: 140048. Published online 2014 Dec 9. https://doi.org/10.1038/sdata.2014.48

Wastewater Treatment, Advanced (Grades 3 and 4)

Treatment Process Monitoring, Evaluation, & Adjustment

86. **a.** Corrosion, staining, chemical burns, difficulty in handling, and vapors.

Wastewater Treatment, Introductory (Grades 1 and 2)

Security, Safety, & Administrative Procedures

Treatment Process Monitoring, Evaluation, & Adjustment

87. **a.** It is the measure of how quickly oxygen can be transferred into liquid.

Ref.: Page 560, (Oxygen Transfer Efficiency), Operation of Wastewater Treatment Plants, Vol. 1, 8th edition.

Wastewater Treatment, Introductory (Grades 1 and 2)

Treatment Process Monitoring, Evaluation, & Adjustment

Equipment Operation

88. **c.** It means that the aeration basin is not getting as much oxygen.

Wastewater Treatment, Introductory (Grades 1 and 2)

Treatment Process Monitoring, Evaluation, & Adjustment

89. **a.** Bubble size, water temperature, and mixing.

Ref.: Page 560, Operation of Wastewater Treatment Plants, Vol. 1, 8th edition.

Wastewater Treatment, Introductory (Grades 1 and 2)

Treatment Process Monitoring, Evaluation, & Adjustment

Equipment Operation

90. **c.** It is a measure of specific bacterial uptake by mixed liquor volatile suspended solids (MLVSS).

Ref.: USEPA draft method 1683 (2001), Specific Oxygen Uptake in Biosolids.
An Introduction to Environmental Engineering Laboratory Experiment #1: Oxygen Utilization Rate (OUR) Test.

Wastewater Treatment, Advanced (Grades 3 and 4)

Treatment Process Monitoring, Evaluation, & Adjustment

91. **d.** It is a possible indication of toxics in the aeration basin.

SOUR describes the amount of oxygen used by the microorganisms to consume 1 g of food and is reported as mg/L of oxygen used per gram of organic material per hour. The specific uptake rate is valuable when comparing one aquatic system with another or if a single system is charted over time. The performance of one aeration basin can be compared with another, or the biological activity in a stream can be studied and compared both above and below a waste outfall. Furthermore, toxic or high organic loads can often be detected before severe deterioration of effluent quality occurs. Changes in the SOUR on effluent samples will indicate changes in loading. The introduction of toxic materials can cause a marked decrease in the biological activity as measured by the MLSS respiration. Dramatic decreases in SOUR can indicate this type of toxicity.

Wastewater Treatment, Advanced (Grades 3 and 4)

Treatment Process Monitoring, Evaluation, & Adjustment

92. **a.** Activated sludge

Ref.: The Pennsylvania State Association of Township Supervisors (PSATS) Gannett Fleming, Inc. Dering Consulting Group Penn State Harrisburg Environmental Training Center, Wastewater Treatment Plant Operator Certification Training, Module 17: The Activated Sludge Process Part III. Revised October 2014.

A contact stabilization plant is a variant of an activated sludge system with shorter detention times.

Wastewater Treatment, Introductory (Grades 1 and 2)

Treatment Process Monitoring, Evaluation, & Adjustment

93. **b.** Both are attached growth reactors that require intermittent dosing or contact with air to provide oxygen to the attached growth.

Ref.: Pages 470-485 and 525-565, Operation of Wastewater Treatment Plants, Vol. 1, 8th edition.

Wastewater Treatment, Introductory (Grades 1 and 2)

Treatment Process Monitoring, Evaluation, & Adjustment

94. **a.** Growth of filamentous organisms

Ref.: Page 211, Operation of Wastewater Treatment Plants, Vol. 1, 8th edition.
Pages 70-78, Operation of Wastewater Treatment Plants, Vol. 2, 7th edition.

Wastewater Treatment, Introductory (Grades 1 and 2)

Treatment Process Monitoring, Evaluation, & Adjustment

95. **d.** All of the above.

Ref.: Page 241, Operation of Wastewater Treatment Plants, Vol. 2, 7th edition.
Page 694, Operation of Wastewater Treatment Plants, Vol. 1, 8th edition.
Environ. Sci. Technol. 2015, 49(2): 1051–1059.

Wastewater Treatment, Introductory (Grades 1 and 2)

Security, Safety, & Administrative Procedures

Treatment Process Monitoring, Evaluation, & Adjustment

96. **d.** Calibrate in accordance with the manufacturer's recommendations and check the installation conditions.

Wastewater Treatment, Introductory (Grades 1 and 2)

Equipment Operation

Equipment Evaluation & Maintenance

97. **b.** Special personal protective equipment and disposal equipment may be needed to handle the liquid mercury seals if they have to be replaced.

Ref.: Pages 830-845, Operation of Wastewater Treatment Plants, Vol. 1, 8th edition.

Wastewater Treatment, Introductory (Grades 1 and 2)

Security, Safety, & Administrative Procedures

Equipment Evaluation & Maintenance

98. **a.** They both improve pollutant removal efficiency with recycling from the secondary clarifier.

Ref.: Pages 466-467, page 521, Section 6.2.10.1 to 6.3, Operation of Wastewater Treatment Plants, Vol. 1, 8th edition.

Wastewater Treatment, Introductory (Grades 1 and 2)

Treatment Process Monitoring, Evaluation, & Adjustment

99. **a.** The wetland is not designed to handle high quantities of biological matter and solids as in a treatment plant and is not as efficient as a wastewater treatment plant in removal of biochemical oxygen demand (BOD) and total suspended solids (TSS).

Ref.: Page 63, Use of Constructed Wetlands for Wastewater Treatment and Recycling.

Wastewater Treatment, Advanced (Grades 3 and 4)

Security, Safety, & Administrative Procedures

Treatment Process Monitoring, Evaluation, & Adjustment

100. **b.** Yes, if there is secondary treatment to remove solids, wetlands are sufficient treatment systems.

Ref.: Ecological Engineering. 2005. The functioning of a wetland system used for polishing effluent from a sewage treatment plant.

Wastewater Treatment, Advanced (Grades 3 and 4)

Security, Safety, & Administrative Procedures

Treatment Process Monitoring, Evaluation, & Adjustment

101. **b.** Exceeding that rate would tend to wash solids near the weirs out of the system.

Wastewater Treatment, Introductory (Grades 1 and 2)

Treatment Process Monitoring, Evaluation, & Adjustment

102. **a.** The membrane system must be cleaned with a citric or other acid solution, and the membranes must be regularly backflushed to prevent solids plugging.

Ref.: Pages 31-34, Operation of Wastewater Treatment Plants, Vol. 1, 8th edition.

Wastewater Treatment, Advanced (Grades 3 and 4)

Treatment Process Monitoring, Evaluation, & Adjustment

103. **d.** All of the above.

Ref.: Gilmour, C.C., Bloom, N.S. A case study of mercury and methylmercury dynamics in a Hg-contaminated municipal wastewater treatment plant. Water Air Soil Pollut 80, 799–803 (1995). https://doi.org/10.1007/BF01189731.
Stacy Wong, 1998. Hartford Courant, "MERCURY USED IN EQUIPMENT AT 6 SEWAGE PLANTS"

Mercury seals were and may still be a way of sealing the rotating arms on a trickling filter. See the reference articles as an example.

Wastewater Treatment, Introductory (Grades 1 and 2)

Treatment Process Monitoring, Evaluation, & Adjustment

104. **b.** Trickling filters

Ref.: Water World, IGR Used to Control Filter Fly Infestation, 1999.

Wastewater Treatment, Introductory (Grades 1 and 2)

Treatment Process Monitoring, Evaluation, & Adjustment

105. **d.** Tank covers

Ref.: Water World, IGR Used to Control Filter Fly Infestation, 1999.

Wastewater Treatment, Introductory (Grades 1 and 2)

Treatment Process Monitoring, Evaluation, & Adjustment

106. **d.** All of the above.

Ref.: Section 6.3, Operation of Wastewater Treatment Plants, Vol. 1, 8th edition.

Wastewater Treatment, Introductory (Grades 1 and 2)

Treatment Process Monitoring, Evaluation, & Adjustment

107. **a.** Media size, media distribution, hydraulic loading, and pore size of the openings.

Ref.: C.S.B. Fitzpatrick, Water Science and Technology, "Media properties and their effect on filter performance and backwashing." 38(6): 105-111.
H. E. Hudson Jr., 1956. Journal AWWA, "Factors Affecting Filtration Rates." 48(9): 1138-1154.

Wastewater Treatment, Advanced (Grades 3 and 4)

Treatment Process Monitoring, Evaluation, & Adjustment

108. **b.** A rotating screen that is used to remove fine particles from incoming water.

Ref.: Science Direct. 2019. Biofuels from Algae. See 2.11 Lakeside Equipment Corporation. Raptor® Micro Strainer. USEPA. Microstraining and Disinfection of Combined Sewer Overflows – Phase III.

Wastewater Treatment, Introductory (Grades 1 and 2)

Treatment Process Monitoring, Evaluation, & Adjustment

Equipment Operation

109. **a.** Screening or filtration removal of very large abrasive particles and conditioning of the water.

Wastewater Treatment, Advanced (Grades 3 and 4)

Treatment Process Monitoring, Evaluation, & Adjustment

110. **a.** Plate and frame, spiral wound, and spaghetti strand.

Ref.: Met-Chem Inc., Membrane Filter Plate.
A. von Bubnoff. 2005. Nature. "Spaghetti strands cleanse water supplies".
Suez. Degremont® Water Handbook. Spiral Wound Modules.

Wastewater Treatment, Advanced (Grades 3 and 4)

Treatment Process Monitoring, Evaluation, & Adjustment

111. **b.** By extremely low (less than 1 mg/L) dissolved oxygen (DO) even when aerators are running at their maximum capacity.

Ref.: Page 263, Operation of Wastewater Treatment Plants, Vol. 1, 8th edition.
Pages 19, 49, Operation of Wastewater Treatment Plants, Vol. 2, 7th edition.

Wastewater Treatment, Introductory (Grades 1 and 2)

Treatment Process Monitoring, Evaluation, & Adjustment

112. **a.** A complete mix treatment system introduces waste at various points along the system treatment train, whereas a plug flow system treats everything in sequential order.

Ref.: Pages 306-307, Operation of Wastewater Treatment Plants, Vol. 1, 8th edition.

Wastewater Treatment, Advanced (Grades 3 and 4)

Treatment Process Monitoring, Evaluation, & Adjustment

113. **b.** Fill, react, settle, decant, and rest.

Ref.: Pages 97-109, Operation of Wastewater Treatment Plants, Vol. 2, 7th edition.

Wastewater Treatment, Introductory (Grades 1 and 2)

Treatment Process Monitoring, Evaluation, & Adjustment

114. **d.** pH is a measure of hydrogen ion concentration on a logarithmic scale.

Ref.: Page 7 (Box), Operation of Wastewater Treatment Plants, Vol. 1, 8th edition.
Pages 405 & 564, Basic Science Concepts and Applications, 4th edition.

pH is a measure of hydrogen ion concentration on a logarithmic scale from 1 to 14.

Wastewater Treatment, Introductory (Grades 1 and 2)

Treatment Process Monitoring, Evaluation, & Adjustment

Laboratory Analysis

115. **a.** Greater than 7

Ref.: Pages 405, 564, Basic Science Concepts and Applications, 4th edition.

Wastewater Treatment, Introductory (Grades 1 and 2)

Treatment Process Monitoring, Evaluation, & Adjustment

Laboratory Analysis

116. **a.** The $KMnO_4$ method gives inconsistent results and is no longer used.

Ref.: Page 541, Operation of Wastewater Treatment Plants, Vol. 2, 7th edition.

Wastewater Treatment, Introductory (Grades 1 and 2)

Laboratory Analysis

117. **a.** >7

Ref.: Pages 405, 564, Basic Science Concepts and Applications, 4th edition.

Wastewater Treatment, Introductory (Grades 1 and 2)

Treatment Process Monitoring, Evaluation, & Adjustment

Laboratory Analysis

118. **d.** Oxidize ammonia to nitrate in the aeration tank if possible by adding more mixed liquor suspended solids (MLSS).

Ref.: Pages 587-588, Operation of Wastewater Treatment Plants, Vol. 1, 8th edition.

Wastewater Treatment, Advanced (Grades 3 and 4)

Treatment Process Monitoring, Evaluation, & Adjustment

119. **d.** N_2, NH_3, NO_3, HNO_3, NO_2

Ref.: Pages 397-399, Basic Science Concepts and Applications, 4th edition.
Paragraph 9.5.3.7, page 866, Operation of Wastewater Treatment Plants, Vol. 1, 8th edition.

Wastewater Treatment, Advanced (Grades 3 and 4)

Treatment Process Monitoring, Evaluation, & Adjustment

Laboratory Analysis

120. **d.** be stored in separate rooms with adequate ventilation.

It is never a good idea to store an oxidizer near a reducing agent. The accidental combination can release substantial quantities of energy.

Wastewater Treatment, Introductory (Grades 1 and 2)

Security, Safety, & Administrative Procedures

Treatment Process Monitoring, Evaluation, & Adjustment

121. **c.** Sulfur dioxide

There is no reaction between sulfur dioxide and phosphorus.

Wastewater Treatment, Advanced (Grades 3 and 4)

Treatment Process Monitoring, Evaluation, & Adjustment

122. **d.** Hydrogen sulfide

Ref.: National Research Council (US) Committee on Acute Exposure Guideline Levels. Acute Exposure Guideline Levels for Selected Airborne Chemicals: Volume 9. Washington (DC): National Academies Press (US); 2010. 4, Hydrogen Sulfide Acute Exposure Guideline Levels.

Dihydrogen monoxide has a formula of H_2O. Hydrogen sulfide (H_2S) is a colorless, flammable gas at ambient temperature and pressure. It has an odor similar to that of rotten eggs and is both an irritant and an asphyxiant. The air odor threshold ranges between 0.008 and 0.13 ppm, and olfactory fatigue may occur at 100 ppm.

Wastewater Treatment, Introductory (Grades 1 and 2)

Treatment Process Monitoring, Evaluation, & Adjustment

123. **b.** Cover the aeration tanks and vent to a tall stack.

Ref.: Design Manual: Odor and Corrosion Control in Sanitary Sewerage Systems and Treatment Plants. US Environmental Protection Agency, Washington, DC, EPA/625/1-85/018.

Wastewater Treatment, Introductory (Grades 1 and 2)

Treatment Process Monitoring, Evaluation, & Adjustment

124. **c.** The SVI of less than 150 is a measure of a good settling sludge.

Ref.: pages 261-270, and 924, Operation of Wastewater Treatment Plants, Vol. 1, 8th edition.
Pages 517-520, Operation of Wastewater Treatment Plants, Vol. 2, 7th edition.

SVIs below 80 indicate very old and poor sludge

Wastewater Treatment, Introductory (Grades 1 and 2)

Treatment Process Monitoring, Evaluation, & Adjustment

125. **d.** A digestion in potassium dichromate followed by colorimetric measurement

Ref.: Pages 890-895, Operation of Wastewater Treatment Plants, Vol. 1, 8th edition.

Wastewater Treatment, Introductory (Grades 1 and 2)

Laboratory Analysis

126. **d.** For the soluble BOD5 test, particulate material is filtered out by a 0.45-micron filter, and the test is run as a normal BOD5 test.

Ref.: Pages 520 and 570, Operation of Wastewater Treatment Plants, Vol. 2, 7th edition.

Answer b is incorrect because filtration does not remove the soluble materials.

Wastewater Treatment, Introductory (Grades 1 and 2)

Laboratory Analysis

127. **c.** The insoluble phosphorus stays with the sludge and can be used for land application.

Wastewater Treatment, Advanced (Grades 3 and 4)

Treatment Process Monitoring, Evaluation, & Adjustment

128. **c.** In an anoxic environment, the ORP is slightly negative (–50 mv), and in an anaerobic environment the ORP is much more negative (even less than –100 mv)

Wastewater Treatment, Advanced (Grades 3 and 4)

Treatment Process Monitoring, Evaluation, & Adjustment

129. **c.** Nematodes, rotifers, stalked ciliates, free swimming ciliates, amoeboids

Ref.: Pages 361-403, Operation of Wastewater Treatment Plants, Vol. 2, 7th edition

Wastewater Treatment, Advanced (Grades 3 and 4)

Laboratory Analysis

130. **b.** Accuracy is the true value of the substance, whereas precision is the repeatability of the answer.

Ref.: LabWrite. Accuracy and Precision. https://labwrite.ncsu.edu/Experimental%20Design/accuracyprecision.htm

Accuracy is the practice of obtaining the correct value. Precision is the process of getting the same answer multiple times (making the same mistake on multiple occasions.)

Wastewater Treatment, Introductory (Grades 1 and 2)

Laboratory Analysis

131. **b.** There is no standard for any solids measurement.

Ref.: Standard Methods for the Examination of Water and Wastewater, 23rd edition.

Wastewater Treatment, Introductory (Grades 1 and 2)

Laboratory Analysis

132. **b.** MLSS is dried at 105°C and MLVSS is ignited at 550°C.

Ref.: Pages 830-845, Operation of Wastewater Treatment Plants, Vol. 1, 8th edition.

Wastewater Treatment, Introductory (Grades 1 and 2)

Laboratory Analysis

133. **c.** Less than 1 mg/L

A 0.45-micron filter is used in the laboratory to remove solids from a sample for measurement of TSS and volatile suspended solids (VSS). Anything passing a 0.42-micron filter would also pass a 0.45 micron filter, so the answer is less than 1 mg/L.

Wastewater Treatment, Advanced (Grades 3 and 4)

Treatment Process Monitoring, Evaluation, & Adjustment

Laboratory Analysis

134. **b.** Mesophilic and thermophyllic: 95–145°F

Ref.: Page 278, Pages 830-845, Operation of Wastewater Treatment Plants, Vol. 1, 8th edition.
Page 151, Pages 830-845, Operation of Wastewater Treatment Plants, Vol. 2, 7th edition.

Wastewater Treatment, Introductory (Grades 1 and 2)

Treatment Process Monitoring, Evaluation, & Adjustment

135. **d.** All of the above

Ref.: Page 151, Pages 830-845, Operation of Wastewater Treatment Plants, Vol. 2, 7th edition.

Wastewater Treatment, Introductory (Grades 1 and 2)

Treatment Process Monitoring, Evaluation, & Adjustment

136. **c.** Thermophilic produces class A biosolids.

Ref.: Page 227, Pages 830-845, Operation of Wastewater Treatment Plants, Vol. 2, 7th edition.
USEPA. 1994. A Plain English Guide to the EPA Part 503 Biosolids Rule.

Wastewater Treatment, Advanced (Grades 3 and 4)

Security, Safety, & Administrative Procedures

Treatment Process Monitoring, Evaluation, & Adjustment

137. **a.** High sludge feed concentrations and acceptable gas production.

Ref.: Pages 148, 151, 211, Operation of Wastewater Treatment Plants, Vol. 2, 7th edition.

Wastewater Treatment, Advanced (Grades 3 and 4)

Treatment Process Monitoring, Evaluation, & Adjustment

138. **a.** Short residence time, class A biosolids, and a high rate of gas production.

Ref.: Pages 148, 151, 211, Operation of Wastewater Treatment Plants, Vol. 2, 7th edition.

Wastewater Treatment, Advanced (Grades 3 and 4)

Treatment Process Monitoring, Evaluation, & Adjustment

139. **a.** Increases solids content in sludge.

Ref.: Ch. 12.74, Operation of Wastewater Treatment Plants, Vol. 2, 7th edition.

Wastewater Treatment, Introductory (Grades 1 and 2)

Equipment Operation

140. **b.** Development of two stage (multi-compartment) anaerobic digesters.

Ref.: https://www.nature.com/articles/s41545-018-0021-y
https://farm-energy.extension.org/types-of-anaerobic-digesters/
UK AD and World Biogas Expo 2020. Two Stage Anaerobic Digestion.

Wastewater Treatment, Introductory (Grades 1 and 2)

Treatment Process Monitoring, Evaluation, & Adjustment

Equipment Operation

141. **b.** Digester gas contains a lot of water and carbon dioxide (CO_2), ammonia (NH_3), and hydrogen sulfide (H_2S), and has a low fuel value if used directly.

Ref.: Page 159, Operation of Wastewater Treatment Plants, Vol. 1, 8th edition.

Digester gas can also contain siloxanes that can ruin an internal combustion engine because they oxidize to silicon oxide (sand).

Wastewater Treatment, Introductory (Grades 1 and 2)

Treatment Process Monitoring, Evaluation, & Adjustment

142. **a.** The filter press dewaters sludge by compressing it between plates, whereas the belt filter compresses and shears the sludge as it wraps around rollers.

Ref.: Ch. 12.74, Operation of Wastewater Treatment Plants, Vol. 2, 7th edition.

Wastewater Treatment, Introductory (Grades 1 and 2)

Equipment Operation

143. **b.** Add polymers that cause the sludge feed solids to increase.

Ref.: Ch. 12.74, Page 223, Operation of Wastewater Treatment Plants, Vol. 2, 7th edition.

Wastewater Treatment, Introductory (Grades 1 and 2)

Equipment Operation

144. **a.** Land spreading to incorporate it into the soils.

Wastewater Treatment, Introductory (Grades 1 and 2)

Treatment Process Monitoring, Evaluation, & Adjustment

145. **c.** Arsenic, mercury, molybdenum, cadmium, and selenium

Ref.: Exhibit 4, page 17. 1994. Publication: EPA/831-B-93-002b, "Land Application of Sewage Sludge"

Wastewater Treatment, Introductory (Grades 1 and 2)

Security, Safety, & Administrative Procedures

Treatment Process Monitoring, Evaluation, & Adjustment

146. **a.** Reducing the volume, storage, and transportation costs.

Ref.: EPA Biosolids Technology Fact Sheet, "Belt filter press", 2018.

Wastewater Treatment, Introductory (Grades 1 and 2)

Treatment Process Monitoring, Evaluation, & Adjustment

147. **d.** At 80°C, and above, the HRT is at least 90 days.

Ref.: Pages 148-155, Operation of Wastewater Treatment Plants, Vol. 2, 7th edition.

Temperatures much above 70°C will sterilize the sludge and stop most, if not all, organic degradation. So, no digestion takes place.

Wastewater Treatment, Advanced (Grades 3 and 4)

Treatment Process Monitoring, Evaluation, & Adjustment

148. **a.** Minimum dissolved oxygen (DO) requirements are 0.4–0.8 mg/L.

Ref.: Pages 20-41, 64, 72, 90, Operation of Wastewater Treatment Plants, Vol. 2, 7th edition.

Wastewater Treatment, Introductory (Grades 1 and 2)

Treatment Process Monitoring, Evaluation, & Adjustment

149. **b.** Siloxanes volatilize in gases and interfere with combustion of the generated gas by turning into glass.

Wastewater Treatment, Introductory (Grades 1 and 2)

Treatment Process Monitoring, Evaluation, & Adjustment

150. **c.** It is an ammonium, magnesium, calcium compound that can create deposits in pipes and valves, as well as in the digester itself.

Ref.: Pages 188-190, Operation of Wastewater Treatment Plants, Vol. 2, 7th edition.

Wastewater Treatment, Introductory (Grades 1 and 2)

Treatment Process Monitoring, Evaluation, & Adjustment

Equipment Evaluation & Maintenance

151. **a.** Acceptable metal concentrations and pathogen reduction

Ref.: Pages 227-228, Operation of Wastewater Treatment Plants, Vol. 2, 7th edition.

Wastewater Treatment, Advanced (Grades 3 and 4)

Treatment Process Monitoring, Evaluation, & Adjustment

152. **a.** Composting, thermal treatment, or lime stabilization

Ref.: Page 227, Operation of Wastewater Treatment Plants, Vol. 2, 7th edition.
USEPA. A Plain English Guide to the EPA Part 503 Biosolids Rule, EPA-832/R-93/003 September 1994.

Wastewater Treatment, Advanced (Grades 3 and 4)

Treatment Process Monitoring, Evaluation, & Adjustment

153. **a.** Liquid sludge disposal is easy to apply but is hauled in tankers, whereas dewatered sludge has higher capital and operating cost and lower hauling costs (trucks versus tankers).

Wastewater Treatment, Introductory (Grades 1 and 2)

Treatment Process Monitoring, Evaluation, & Adjustment

154. **a.** analyzed and applied at agronomic rates so that plants will take up excess nutrients.

Ref.: Pages 227-228, Operation of Wastewater Treatment Plants, Vol. 2, 7th edition.

Wastewater Treatment, Introductory (Grades 1 and 2)

Treatment Process Monitoring, Evaluation, & Adjustment

155. **a.** Filter presses produce a drier sludge cake.

Ref.: USEPA Fact Sheet, "Recessed plate presses", 2018.

Wastewater Treatment, Introductory (Grades 1 and 2)

Treatment Process Monitoring, Evaluation, & Adjustment

Equipment Operation

156. **a.** The organic matter is hydrolyzed and made soluble and then fermented into acetic acids, hydrogen, and carbon dioxide (CO_2), and then into methane.

Ref.: Section 12.01, Operation of Wastewater Treatment Plants, Vol. 2, 7th edition.

This is part of the process description of anaerobic degradation.

Wastewater Treatment, Advanced (Grades 3 and 4)

Treatment Process Monitoring, Evaluation, & Adjustment

157. **c.** a wet sludge, gas ($CH_4 + CO_2$), acids, and liquid.

Ref.: Figure 12.2, Operation of Wastewater Treatment Plants, Vol. 2, 7th edition.

Wastewater Treatment, Advanced (Grades 3 and 4)

Treatment Process Monitoring, Evaluation, & Adjustment

158. **c.** both acid formers and methanogens

Ref.: Section 12, Operation of Wastewater Treatment Plants, Vol. 2, 7th edition.

Wastewater Treatment, Advanced (Grades 3 and 4)

Treatment Process Monitoring, Evaluation, & Adjustment

159. **b.** 6.5–7.5

Ref.: Pages 183-84 (see stuck digesters), Operation of Wastewater Treatment Plants, Vol. 2, 7th edition.

Anaerobic digestion is governed principally by the methanogens. If the pH range gets outside of the range of the methanogens, gas production will not occur, creating a "stuck digester."

Wastewater Treatment, Advanced (Grades 3 and 4)

Treatment Process Monitoring, Evaluation, & Adjustment

160. **c.** Bacterial conversion of volatile organics into organic acids (e.g., acetic acid) and propionic acid, and then methane formers convert the acids into methane and CO_2.

Ref.: Pages 150-155, Operation of Wastewater Treatment Plants, Vol. 2, 7th edition.

Wastewater Treatment, Advanced (Grades 3 and 4)

Treatment Process Monitoring, Evaluation, & Adjustment

161. **b.** Too much new sludge fed to the digester beyond what the prokaryotic organisms can use, causing the pH to drop.

Ref.: Pages 183-184, Operation of Wastewater Treatment Plants, Vol. 2, 7th edition.

Wastewater Treatment, Advanced (Grades 3 and 4)

Treatment Process Monitoring, Evaluation, & Adjustment

162. **b.** VFAs are the food that methanogens (PAOs) use to produce methane and carbon dioxide (CO_2) and they accumulate phosphorus.

Ref.: Pages 211, 523, 150-155, Operation of Wastewater Treatment Plants, Vol. 2, 7th edition.

Wastewater Treatment, Advanced (Grades 3 and 4)

Treatment Process Monitoring, Evaluation, & Adjustment

163. **c.** Methanogenic inactivation caused by low pH.

Ref.: Pages 211, 523, Operation of Wastewater Treatment Plants, Vol. 2, 7th edition.

Too much volatile fatty acid (VFA) production results in a "sour" digester and the pH decreases; therefore, methanogens will not convert VFAs to methane.

Wastewater Treatment, Advanced (Grades 3 and 4)

Treatment Process Monitoring, Evaluation, & Adjustment

164. **a.** Add calcium carbonate and lime carefully to raise the pH.

Wastewater Treatment, Advanced (Grades 3 and 4)

Treatment Process Monitoring, Evaluation, & Adjustment

165. **a.** Put it back in the head end of the plant for further processing.

Wastewater Treatment, Advanced (Grades 3 and 4)

Treatment Process Monitoring, Evaluation, & Adjustment

166. **b.** The pH of the limed sludge must be raised to 10–12.

Ref.: https://pdfs.semanticscholar.org/4776/7d71d1fcb8254aabe57cc4f5dadbf819dd57.pdf. US Patent number 4,306,978. Method for Lime Stabilization of Wastewater Treatment Plant Sludges.
USEPA. EPA-625/1-82/014. Design Manual Dewatering Municipal Wastewater Sludges.

Wastewater Treatment, Introductory (Grades 1 and 2)

Treatment Process Monitoring, Evaluation, & Adjustment

167. **b.** A method of creating a soil amendment through aerobic degradation.

Ref.: USEPA, Beltsville Aerated Pile Manual. U.S. Environmental Protection Agency, Washington, D.C., EPA/600/8-80/022.

Wastewater Treatment, Advanced (Grades 3 and 4)

Treatment Process Monitoring, Evaluation, & Adjustment

168. **a.** A manhole is a confined space as defined by the Occupational Safety and Health Administration (OSHA.

Ref.: Pages 265-270, and 731-732, Operation of Wastewater Treatment Plants, Vol.2, 7th edition. Confined Spaces, OSHA Regulations. https://ohsonline.com/articles/2018/08/01/we-must-change-the-statistics-of-confined-space-injuries-and-fatalities.aspx

It is recommended that one read the OSHA definition of a confined space because many employees die each year from entering confined space.

Wastewater Treatment, Introductory (Grades 1 and 2)

Security, Safety, & Administrative Procedures

169. **b.** The density of aerated wastewater is less than the density of the human body, and it is impossible to swim out.

Ref.: OSHA regulations: Standard number: 1926.106 - Working over or near water.

Wastewater Treatment, Introductory (Grades 1 and 2)

Security, Safety, & Administrative Procedures

170. **b.** For worker protection from infection, spills, and splashes.

Ref.: Ch. 16.210 (laboratory safety), Operation of Wastewater Treatment Plants, Vol. 2, 7th edition.

Wastewater Treatment, Introductory (Grades 1 and 2)

Security, Safety, & Administrative Procedures

171. **b.** In separate spill proof and/or flammables cabinets.

Ref.: SHA Regulations: Standard Number: 1910.1450 App A - National Research Council Recommendations Concerning Chemical Hygiene in Laboratories (Non-Mandatory)

leaks of caustics and acids when combined can release significant quantities of gas and energy.

Wastewater Treatment, Introductory (Grades 1 and 2)

Security, Safety, & Administrative Procedures

Laboratory Analysis

172. **d.** Municipal workers are exempt, but outside contractors are fully covered by the regulations.

Ref.: Under paragraph 5 in the definitions of the Occupational Safety and Health Act of 1970 (that is the existing OSHA law). There is a specific exemption for states and any subdivisions thereof, which includes municipalities, governmental agencies, public universities, and municipalities and their operations.

All government agencies are exempt from compliance with OSHA regulations by statute.

Wastewater Treatment, Introductory (Grades 1 and 2)

Security, Safety, & Administrative Procedures

173. **b.** In the event of a discharge of oil, the discharge should be reported to the state Environmental Protection Agency (EPA) and USEPA within 24 h.

Ref.: USEPA. Oil Spills Prevention and Preparedness Regulations.

Wastewater Treatment, Introductory (Grades 1 and 2)

Security, Safety, & Administrative Procedures

174. **c.** Employees should be alerted to vapor/fire/explosion hazards and not to smoke because the fuel will evaporate in the aeration basin.

Wastewater Treatment, Advanced (Grades 3 and 4)

Wastewater Treatment, Introductory (Grades 1 and 2)

Security, Safety, & Administrative Procedures

175. **a.** Close the inlet and outlet valves to stop the flow through the pump volute.

The answer depends on the motor design. Safety is the consideration, and a rotating motor shaft can generate electrical currents. The motor can act as a generator if the impeller is spinning.

Wastewater Treatment, Introductory (Grades 1 and 2)

Equipment Evaluation & Maintenance

176. **a.** It must be reported within 24 h verbally to the state or federal government, and a written report must be filed within five days. The report must include an analysis of the causes and measures taken to prevent recurrence.

Ref.: Water Online, 2000. "Clean Water Act Prohibits Sewage 'Bypasses'"

Wastewater Treatment, Introductory (Grades 1 and 2)

Security, Safety, & Administrative Procedures

Treatment Process Monitoring, Evaluation, & Adjustment

177. **c.** Everything listed in (b) plus the required sample volume and the protocols for split sampling with the industry.

Ref.: Pages 821-824, 814-815, and 825-827, Operation of Wastewater Treatment Plants, Vol. 1, 8th edition.

Wastewater Treatment, Advanced (Grades 3 and 4)

Security, Safety, & Administrative Procedures

Treatment Process Monitoring, Evaluation, & Adjustment

178. **b.** To kill any remaining bacteria and spore-forming organisms.

Ref.: Section 1.4.2.5, Operation of Wastewater Treatment Plants, Vol. 1, 8th edition.

Wastewater Treatment, Introductory (Grades 1 and 2)

Treatment Process Monitoring, Evaluation, & Adjustment

179. **c.** 15 min

Ref.: Ten States Standards: Disinfection Paragraph 102.44 (2014 edition)

The Ten States Standards recommends a minimum of 15 min detention time at maximum hourly flow rate with a minimum chlorine dosage of 6 mg/L, depending on the wastewater treatment process.

Wastewater Treatment, Introductory (Grades 1 and 2)

Treatment Process Monitoring, Evaluation, & Adjustment

180. **b.** $HClO^{-}$, $OCl^{=}$

Ref.: Pages 28-30 and Chapter 7, pages 581-584, Operation of Wastewater Treatment Plants, Vol. 1, 8th edition.

Wastewater Treatment, Introductory (Grades 1 and 2)

Treatment Process Monitoring, Evaluation, & Adjustment

181. **c.** Application of chlorine until it oxidizes all amines and is present as free chlorine.

Ref.: Page 586, Operation of Wastewater Treatment Plants, Vol. 1, 8th edition.

Wastewater Treatment, Introductory (Grades 1 and 2)

Treatment Process Monitoring, Evaluation, & Adjustment

182. **b.** Aquatic toxicity of chlorine is 37 mg/L for many species.

Ref.: Page 640, Operation of Wastewater Treatment Plants, Vol. 1, 6th edition.

Wastewater Treatment, Introductory (Grades 1 and 2)

Treatment Process Monitoring, Evaluation, & Adjustment

183. **b.** Sulfur dioxide and sodium bisulfite

Ref.: Page 640, Operation of Wastewater Treatment Plants, Vol. 1, 8th edition.

Wastewater Treatment, Introductory (Grades 1 and 2)

Treatment Process Monitoring, Evaluation, & Adjustment

184. **a.** Disinfection depends on transmittance of the fluid, intensity of radiation, and time of exposure. Even if properly treated, organisms can still regrow.

Ref.: Pages 662-685 and Ch. 7.10, Operation of Wastewater Treatment Plants, Vol. 2, 7th edition.

Wastewater Treatment, Advanced (Grades 3 and 4)

Wastewater Treatment, Introductory (Grades 1 and 2)

Treatment Process Monitoring, Evaluation, & Adjustment

185. **b.** Chlorine is a toxic gas that is difficult to transport and can represent a danger to the community when it is stored in bulk or accidentally released or mishandled.

Ref.: Pages 580-81, Operation of Wastewater Treatment Plants, Vol. 1, 8th edition.

Wastewater Treatment, Advanced (Grades 3 and 4)

Wastewater Treatment, Introductory (Grades 1 and 2)

Security, Safety, & Administrative Procedures

Treatment Process Monitoring, Evaluation, & Adjustment

186. **b.** It requires safe handling and special personal protective equipment (PPE).

Ref.: Ch. 7.3.2.2, Operation of Wastewater Treatment Plants, Vol. 1, 8th edition.

Wastewater Treatment, Advanced (Grades 3 and 4)

Wastewater Treatment, Introductory (Grades 1 and 2)

Security, Safety, & Administrative Procedures

Treatment Process Monitoring, Evaluation, & Adjustment

187. **a.** It is a hazardous gas that cannot be stored, must be generated on site, and is highly corrosive as an oxidizer.

Ref.: Pages 579, and 685-87, Operation of Wastewater Treatment Plants, Vol. 1, 8th edition.

Wastewater Treatment, Advanced (Grades 3 and 4)

Wastewater Treatment, Introductory (Grades 1 and 2)

Security, Safety, & Administrative Procedures

Treatment Process Monitoring, Evaluation, & Adjustment

188. **d.** concentration, contact time, and dose

Ref.: Paragraphs 7.3.4 and 7.10, Operation of Wastewater Treatment Plants, Vol. 1, 8th edition.

Wastewater Treatment, Introductory (Grades 1 and 2)

Treatment Process Monitoring, Evaluation, & Adjustment

189. **a.** Bacterial spores

Ref.: Rahn, O. 1945. "Physical methods of sterilization of micro-organisms. Bacteriological Reviews 9, pp.1–7. As quoted in Practical Wastewater Treatment, Op. Cit.

Wastewater Treatment, Introductory (Grades 1 and 2)

Treatment Process Monitoring, Evaluation, & Adjustment

190. **c.** Ozone

Ref.: Paragraph 7.10, Operation of Wastewater Treatment Plants, Vol. 1, 8th edition.

Wastewater Treatment, Advanced (Grades 3 and 4)

Wastewater Treatment, Introductory (Grades 1 and 2)

Treatment Process Monitoring, Evaluation, & Adjustment

191. **c.** Ozone does not leave a residual in the water, and it is easy to re-contaminate the water supply.

Ref.: Paragraph 7.10, Operation of Wastewater Treatment Plants, Vol. 1, 8th edition.

Wastewater Treatment, Advanced (Grades 3 and 4)

Wastewater Treatment, Introductory (Grades 1 and 2)

Treatment Process Monitoring, Evaluation, & Adjustment

192. **a.** A very effective disinfecting compound made by reacting acetic acid and hydrogen peroxide.

Wastewater Treatment, Advanced (Grades 3 and 4)

Treatment Process Monitoring, Evaluation, & Adjustment

193. **a.** It doesn't generate disinfection byproducts.

Ref.: Final Report: Disinfection Pilot Trial for Little Miami WWTP, Cincinnati, OH, Oct. 17, 2016; Page 4 of 27

Wastewater Treatment, Advanced (Grades 3 and 4)

Treatment Process Monitoring, Evaluation, & Adjustment

194. **b.** A byproduct of the reaction between chlorine and ammonia.

Ref.: page 1038, Operation of Wastewater Treatment Plants, Vol. 1, 8th edition.

Wastewater Treatment, Introductory (Grades 1 and 2)

Treatment Process Monitoring, Evaluation, & Adjustment

195. **a.** The application of excess chlorine to reduce all ammonia to nitrogen gas (N_2).

Ref.: Page 586, Section 7.3, Operation of Wastewater Treatment Plants, Vol. 1, 8th edition.

Wastewater Treatment, Advanced (Grades 3 and 4)

Wastewater Treatment, Introductory (Grades 1 and 2)

Treatment Process Monitoring, Evaluation, & Adjustment

196. **a.** A flow measurement device (a flowmeter).

Ref.: Pages 318-330, Basic Science Concepts and Applications, 4th edition.

Wastewater Treatment, Introductory (Grades 1 and 2)

Treatment Process Monitoring, Evaluation, & Adjustment

Equipment Operation

197. **a.** Parshall flume and variations

Ref.: Pages 305, 315-335, Basic Science Concepts and Applications, 4th edition.
Page 20, Operation of Wastewater Treatment Plants, Vol. 1, 8th edition.

Wastewater Treatment, Introductory (Grades 1 and 2)

Treatment Process Monitoring, Evaluation, & Adjustment

Equipment Operation

198. **b.** If properly installed, it is an accurate measuring device for open channels.

Ref.: Pages 305, 325-329, Basic Science Concepts and Applications, 4th edition.

Wastewater Treatment, Advanced (Grades 3 and 4)

Wastewater Treatment, Introductory (Grades 1 and 2)

Treatment Process Monitoring, Evaluation, & Adjustment

Equipment Operation

199. **a.** Palmer-Bowlus flumes

The Palmer-Bowlus flume is designed as a modified type of Parshall flowmeter designed for inserting into existing sewer lines because of its wide open throat and ability to pass solids.

Wastewater Treatment, Advanced (Grades 3 and 4)

Wastewater Treatment, Introductory (Grades 1 and 2)

Treatment Process Monitoring, Evaluation, & Adjustment

Equipment Operation

Equipment Evaluation & Maintenance

200. **a.** Uniform flow upstream of the flowmeter, where the uniform flow is quiescent.

Ref.: Ch. 3, US Bureau of Reclamation, Water Measurement Manual

Wastewater Treatment, Advanced (Grades 3 and 4)

Wastewater Treatment, Introductory (Grades 1 and 2)

Equipment Operation

Equipment Evaluation & Maintenance

201. **c.** 2–5 percent

Ref.: Ch. 3, US Bureau of Reclamation, Water Measurement Manual

Most electronic flowmeters have an accuracy of about 2 percent. The open channel flowmeter has a general accuracy of greater than 10 percent if good installation practices are followed.

Wastewater Treatment, Advanced (Grades 3 and 4)

Wastewater Treatment, Introductory (Grades 1 and 2)

Treatment Process Monitoring, Evaluation, & Adjustment

Equipment Operation

Equipment Evaluation & Maintenance

202. **c.** Magnetic flowmeter (magmeter)

Magnetic flowmeters are one of the preferred options because of their insensitivity to solids.

Wastewater Treatment, Advanced (Grades 3 and 4)

Wastewater Treatment, Introductory (Grades 1 and 2)

Equipment Operation

Equipment Evaluation & Maintenance

203. **b.** Pipe partially full

Without a full pipe, flowmeters will not give an accurate measurement.

Wastewater Treatment, Advanced (Grades 3 and 4)

Wastewater Treatment, Introductory (Grades 1 and 2)

Treatment Process Monitoring, Evaluation, & Adjustment

Equipment Operation

204. **c.** A way of using compressed air in the bottom of a pipe as a pump to move water.

Wastewater Treatment, Introductory (Grades 1 and 2)

Treatment Process Monitoring, Evaluation, & Adjustment

Equipment Operation

205. **a.** A plunger style pump that is often used to pump sludge.

Ref.: Pages 263-304, Basic Science Concepts and Applications, 4th edition.
Pages 432-33. Operation of Wastewater Treatment Plants, Vol. 1, 8th edition.

Wastewater Treatment, Introductory (Grades 1 and 2)

Equipment Operation

Equipment Evaluation & Maintenance

206. **a.** A positive displacement blower that uses figure-8-shaped rotating impellers to provide positive displacement and move air.

Ref.: Pages 273, 432-433, Operation of Wastewater Treatment Plants, Vol. 1, 8th edition.

Wastewater Treatment, Introductory (Grades 1 and 2)

Equipment Operation

Equipment Evaluation & Maintenance

207. **c.** An ancient way of moving water from one level to a higher level.

Ref.: Page 354, Figure 15.11, Operation of Wastewater Treatment Plants, Vol. 2, 7th edition.

Wastewater Treatment, Advanced (Grades 3 and 4)

Wastewater Treatment, Introductory (Grades 1 and 2)

Equipment Operation

Equipment Evaluation & Maintenance

208. **a.** Centrifugal blowers' and axial compressors' capacities decrease markedly with increasing pressure, whereas a positive displacement blower's capacity is relatively independent of pressure.

Wastewater Treatment, Advanced (Grades 3 and 4)

Wastewater Treatment, Introductory (Grades 1 and 2)

Equipment Operation

Equipment Evaluation & Maintenance

209. **c.** The pressure against which there is no more flow through the blower

Ref.: Pages 371, and 429-431, Operation of Wastewater Treatment Plants, Vol. 1, 8th edition.

Wastewater Treatment, Introductory (Grades 1 and 2)

Equipment Operation

Equipment Evaluation & Maintenance

210. **c.** The pump can overheat and turn the water into steam, damaging to the pump.

Ref.: AICHE Library, "Case histories of pump explosions while running isolated", 2001, 20(2); 152-156.

Wastewater Treatment, Introductory (Grades 1 and 2)

Security, Safety, & Administrative Procedures

Equipment Operation

Equipment Evaluation & Maintenance

211. **c.** The pump motor could burn out because it is overloaded.

Ref.: Accendo Reliability, "Don't Run Centrifugal Pumps on the End of the Curve."

Wastewater Treatment, Introductory (Grades 1 and 2)

Equipment Operation

Equipment Evaluation & Maintenance

212. **d.** All of the above.

Ref.: Pages 400-402, Operation of Wastewater Treatment Plants, Vol. 2, 7th edition.

Wastewater Treatment, Introductory (Grades 1 and 2)

Equipment Operation

Equipment Evaluation & Maintenance

213. **a.** A split ring that enables one to tighten the packing on the pump to prevent leaks around the shaft.

Ref.: Ch. 15, Pages 340-364, Operation of Wastewater Treatment Plants, Vol. 2, 7th edition.

Wastewater Treatment, Introductory (Grades 1 and 2)

Equipment Operation

Equipment Evaluation & Maintenance

214. **c.** It is used to flush the pump seals from the outside, keeping them clear of grit and dirt, which could shorten their life.

Ref.: Ch. 15, pages 340-364, Operation of Wastewater Treatment Plants, Vol. 2, 7th edition.

Wastewater Treatment, Introductory (Grades 1 and 2)

Equipment Operation

Equipment Evaluation & Maintenance

215. **b.** be allowed to run dry.

A submersible pump generally has a self-contained motor, and the liquid that surrounds the motor helps cool it and prevents overheating.

Wastewater Treatment, Introductory (Grades 1 and 2)

Equipment Operation

216. **a.** pay attention to the direction of rotation on the pump casing, the electrical connections, and the voltages.

Ref.: Ch. 15, Operation of Wastewater Treatment Plants, Vol. 2, 7th edition.

Wastewater Treatment, Introductory (Grades 1 and 2)

Equipment Operation

Equipment Evaluation & Maintenance

217. **b.** Power demand increases with the cube of the RPM ratios.

Ref.: Section 7.8, page 996, Operation of Wastewater Treatment Plants, Vol. 1, 8th edition.

Wastewater Treatment, Introductory (Grades 1 and 2)

Equipment Operation

Equipment Evaluation & Maintenance

218. **c.** Closing a valve or stopping a pump can cause a transient pressure spike upstream and downstream of the system.

Ref.: Page 337, Basic Science Concepts and Applications, 4th edition.

The sudden change in flow can create sudden high-pressure or low-pressure surges, which can damage the pump or cause water hammer. In extreme cases, water hammer can cause a blow out of valves and pipelines.

Wastewater Treatment, Introductory (Grades 1 and 2)

Treatment Process Monitoring, Evaluation, & Adjustment

Equipment Operation

Equipment Evaluation & Maintenance

219. **c.** The sum of the friction losses, the suction lift, the entrance and discharge losses on the piping.

Ref.: Page 990, Appendix 7.6, Operation of Wastewater Treatment Plants, Vol. 1, 8th edition.

Wastewater Treatment, Introductory (Grades 1 and 2)

Treatment Process Monitoring, Evaluation, & Adjustment

Equipment Operation

Equipment Evaluation & Maintenance

220. **d.** Atmospheric pressure and elevation

A perfect vacuum can only pull water to a maximum of 33.9 ft at sea level because that is the force of the weight of air pushing against the vacuum. With increasing elevation, atmospheric pressure decreases and so does the maximum suction lift.

Wastewater Treatment, Introductory (Grades 1 and 2)

Treatment Process Monitoring, Evaluation, & Adjustment

Equipment Operation

221. **a.** The difference between atmospheric pressure and the lift and dynamic head losses on the suction side of a pump.

Ref.: Pages 274-277, Basic Science Concepts and Applications, 4th edition.

Wastewater Treatment, Introductory (Grades 1 and 2)

Treatment Process Monitoring, Evaluation, & Adjustment

Equipment Operation

Equipment Evaluation & Maintenance

222. **a.** The maximum pressure a centrifugal pump can produce at zero flow.

Ref.: Pages 376-78, 990-1000, Operation of Wastewater Treatment Plants, Vol. 1, 8th edition.Pages 297-304, Basic Science Concepts and Applications, 4th edition.

Wastewater Treatment, Introductory (Grades 1 and 2)

Treatment Process Monitoring, Evaluation, & Adjustment

Equipment Operation

Equipment Evaluation & Maintenance

223. **b.** creating pressure surges, which can damage other pumps in the system.

Ref.: Pages 337, 341 & 569, Basic Science Concepts and Applications, 4th edition

Wastewater Treatment, Introductory (Grades 1 and 2)

Treatment Process Monitoring, Evaluation, & Adjustment

Equipment Operation

224. **a.** there are rags and clogs in the flow.

Wastewater Treatment, Introductory (Grades 1 and 2)

Equipment Operation

225. **a.** has a recessed impeller design.

Wastewater Treatment, Introductory (Grades 1 and 2)

Equipment Operation

226. **a.** Viscous liquids and chemical slurries, corrosive chemicals, sludge, and mud.

Wastewater Treatment, Introductory (Grades 1 and 2)

Treatment Process Monitoring, Evaluation, & Adjustment

Equipment Operation

227. **a.** Bacteria

Wastewater Treatment, Advanced (Grades 3 and 4)

Wastewater Treatment, Introductory (Grades 1 and 2)

Treatment Process Monitoring, Evaluation, & Adjustment

Laboratory Analysis

228. **d.** Flagellates

Wastewater Treatment, Advanced (Grades 3 and 4)

Wastewater Treatment, Introductory (Grades 1 and 2)

Treatment Process Monitoring, Evaluation, & Adjustment

Laboratory Analysis

229. **c.** Stalked ciliates

Wastewater Treatment, Advanced (Grades 3 and 4)

Wastewater Treatment, Introductory (Grades 1 and 2)

Treatment Process Monitoring, Evaluation, & Adjustment

Laboratory Analysis

230. **d.** Filaments

Wastewater Treatment, Advanced (Grades 3 and 4)

Wastewater Treatment, Introductory (Grades 1 and 2)

Treatment Process Monitoring, Evaluation, & Adjustment

Laboratory Analysis

231. **b.** Rotifers

Wastewater Treatment, Advanced (Grades 3 and 4)

Wastewater Treatment, Introductory (Grades 1 and 2)

Treatment Process Monitoring, Evaluation, & Adjustment

Laboratory Analysis

Appendix A: Common Conversion Factors

Length

1 inch (in.) = 2.54 centimeters (cm)

100 cm = 1 meter (m)

1 m = 39.37 in.

1 m = 3.281 feet (ft)

1 yard = 0.9144 m

1,000 m = 1 kilometer (km)

1 km = 1.609 miles (mi)

Area

1 acre (ac) = 43,560 square feet (ft^2)

1 acre-foot (acre-ft) = 43,560 cubic feet (ft^3)

2.4711 ac = 1 hectare (ha)

1 ha = 0.4047 ac

1 ha = 10,000 square meters (m^2)

1 square mile (mi^2) = 640 ac

Volume

2 pints (pt) = 1 quart (qt)

8 pt = 1 gallon (gal)

4 qt = 1 gal

1 qt = 32 fl oz

1 gal = 128 fl oz

1 gal = 3.785 liters (L)

1 L = 1.0567 qt

1 L = 1,000 milliliters (mL)

3,785 mL = 1 gal

1,000 L = 1 cubic meter (m^3)

1 mL = 1 cubic centimeter (cm^3)

1 = 35.3 cubic feet (ft^3)

1 ft^3 = 7.48 gal

1 million gallons (mil gal) = 3.07 acre-ft

1 acre-ft = 325,829 gal

Weight

1 gram (g) = 1,000 milligrams (mg)

1,000 g = 1 kilogram (kg)

1 pound (lb) = 454 gm

1 lb = 7,000 grains (gr)

1 kg = 2.205 pounds (lb)

2,000 lb = 1 ton

1 mg/L = 1 part per million (ppm)

1 grain per gal (gpg) = 17.1 ppm

Density

Water has a density of 1 g per mL (1 g/mL) or 8.34 lb/gal or 62.4 lb/ft^3

Pressure

1 pound per square inch (psi) = 2.307 ft of water

1 ft of water = 0.4335 psi

1 atmosphere (atm) = 14.7 psi

1 atm = 29.92 in. of mercury

1 atm = 33.90 ft of water

1 atm = 760 millimeters (mm) of mercury

Power

1 horsepower (hp) = 0.746 kilowatts (kW)

1 kW = 1.341 hp

Flow

1 miner's inch = 1.5 ft^3/min

1 ft^3 = 448.8 gal/min (gpm)

1 ft^3/s = 0.6463 million gallons per day (mgd)

1 mgd = 1.547 ft^3/s

Concentration

1% solution = 1 part in 100 parts

1% solution = 10,000 parts per million (ppm)

1 ppm = 1 mg/L

1 grain per gal (gpg) = 17.12 ppm

Temperature

Degrees Fahrenheit (°F) = (9°F/5°C)(°C) + 32°F

Degrees Celsius (°C) = (°F – 32°F)(5°C/9°F)

Appendix B: Summary of Wastewater Treatment Equations (All Grades)

Area Equations

Area of a rectangle = (Length)(Width)

Area of a circle (tank) = (0.785)(Diameter)^2 or πr^2

Area of a parallelogram = (Base)(Height)

$$\text{Area of a trapezoid} = (\text{Altitude})\frac{(\text{Base}_1 + \text{Base}_2)}{2}$$

Basic Electricity Formulas

Voltage = (amps)(Resistance, ohms)

Resistance, ohms = Voltage/amps

$$\text{Current process reading} = \frac{(\text{Live signal, mA} - 4\text{ mA offset})(\text{Maximum capacity})}{16\text{ milliamp span}}$$

Biochemical Oxygen Demand Loading Equations

BOD_5, lb/d = (BOD_5, mg/L)(Number of mgd)(8.34 lb/gal)

Biochemical Oxygen Demand Unseeded and Seeded Equations

$$BOD_5\text{ unseeded, mg/L} = \frac{(\text{Initial DO, mg/L} - \text{Final DO, mg/L})(\text{Total volume, mL})}{\text{Sample volume, mL}}$$

$$\text{Seed correction, mg/L} = \frac{(BOD_5\text{ of seed stock, mg/L})(\text{Seed stock, mg/L})}{\text{Total volume, mL}}$$

$$BOD_5\text{ seeded, mg/L} = \frac{(\text{Initial DO, mg/L} - \text{Final DO, mg/L} - \text{Seed correction, mg/L})(\text{Total volume, mL})}{\text{Sample volume, mL}}$$

Biosolids Concentration Factor

$$\text{CF} = \frac{\text{Percent thickened biosolids}}{\text{Percent influent biosolids}}$$

Biosolids Pumping and Production Formulas

$$\text{Estimated pumping rate} = \frac{(\text{Influent TSS, mg/L} - \text{Effluent TSS, mg/L})(\text{Flow, mgd})(8.34\text{ lb/gal})}{(\text{Percent solids in sludge})(\text{Sludge, lb/gal})(1{,}440\text{ min/d})}$$

$$\text{Biosolids, lb/mil gal} = \frac{(\text{Biosolids, gal})(8.34\text{ lb/gal})}{(\text{Flow, mgd})(\text{Number of days})}$$

$$\text{Biosolids, lb/mil gal} = \frac{(\text{Biosolids, gal/d})(8.34\text{ lb/gal})}{\text{Flow, mil gal}}$$

$$\text{Biosolids, wet tons/yr} = \frac{(\text{Biosolids, lb/mil gal})(\text{mgd})(365\text{ d/yr})}{2{,}000\text{ lb/ton}}$$

$$\text{Estimated pumping rate} = \frac{(\text{Influent TSS, mg/L} - \text{Effluent TSS, mg/L})(\text{Flow, mgd})(8.34\text{ lb/gal})}{(\text{Percent solids in sludge})(\text{Sludge, lb/gal})(1{,}440\text{ min/d})}$$

Biosolids Retention Time Equation

$$\text{BRT, days} = \frac{\text{Digester working volume, gal}}{\text{Influent flow, gpd}}$$

Biosolids Volume Index and Biosolids Density Index Equations

$$\text{BVI} = \frac{(\text{Settled biosolids, mL/L})(1{,}000\ \text{mg/g})}{\text{MLSS, mg/L}}$$

$$\text{BDI} = \frac{(\text{MLSS, mg/L})(100)}{(\text{Settled biosolids, mL/L})(1{,}000\ \text{mg/g})}$$

Centrifuge Thickening Equation

Hydraulic loading, gpd = (Sludge flow, gpm)(1,440 min/d)

$$\text{Feed time, min} = \frac{(\text{Capacity, ft}^3)(\text{Solids, \%/100\%})(7.48\ \text{gal/ft}^3)(8.34\ \text{lb/gal})}{(\text{Flow, gpm})(\text{Solids concentration, \%/100\%})(8.34\ \text{lb/gal})}$$

or simplified:

$$\text{Feed time, min} = \frac{(\text{Capacity, ft}^3)(\text{Solids, \%})(7.48\ \text{gal/ft}^3)}{(\text{Flow, gpm})(\text{Solids concentration, \%})}$$

Chemical Feed Solution Settings

$$\text{Feed rate, mL/min} = \frac{(\text{gpd})(3{,}785\ \text{mL/gal})}{1{,}440\ \text{min/d}} \quad \text{or}$$

$$\text{Number of mL/min} = \frac{(\text{Number of gallons used})(3{,}785\ \text{mL/gal})}{1{,}440\ \text{min/d}}$$

Chemical Oxygen Demand Loading Formula

COD, lb/d = (COD, mg/L)(Number of mgd)(8.34 lb/gal)

Chemistry and Laboratory Equations

$$\text{Moles} = \frac{\text{Grams of chemical}}{\text{Gram formula weight}}$$

$$\text{Percent of element in compound} = \frac{(\text{Molecular Wt of the element})(100\%)}{\text{Molecular Wt of compound}}$$

$$\text{BOD}_5 \text{ unseeded, mg/L} = \frac{(\text{Initial DO, mg/L} - \text{Final DO, mg/L})(\text{Total volume, mL})}{\text{Sample volume, mL}}$$

$$\text{Seed correction, mg/L} = \frac{(\text{BOD}_5 \text{ of seed stock, mg/L})(\text{Seed stock, mg/L})}{\text{Total volume, mL}}$$

BOD_5 seeded, mg/L =

$$\frac{(\text{Initial DO, mg/L} - \text{Final DO, mg/L} - \text{Seed correction, mg/L})(\text{Total volume, mL})}{\text{Sample volume, mL}}$$

$$\text{Percent of an element in a compound} = \frac{(\text{Molecular Wt of the element})(100\%)}{\text{Molecular Wt of the compound}}$$

$$\text{Molarity} = \frac{\text{Moles solute}}{\text{Liters solution}}$$

$$\text{Normality } (N) = \frac{\text{Number of gram-equivalents of Solute}}{\text{Number of liters of Solution}}$$

$$\text{Dosage, mg/L} = \frac{(\text{Stock, mL})(1{,}000 \text{ mg/gram})(\text{Concentration in g/L})}{\text{Sample size, mL}}$$

$$\text{Percent VS} = \frac{(\text{Solids lost, g})(100\%)}{\text{Weight of total solids, g}}$$

$$\text{Alum, mg/L} = \frac{(\text{Stock, mL})(1{,}000 \text{ mg/g})(\text{Concentration in g/L})}{\text{Sample size, mL}}$$

Solids (ash), g = Sample and dish dried, g – Burnt sample, g

Circumference Formulas

Circumference = π(Diameter)

Circumference = 2π(radius) or $2\pi r$

Common Conversion Factors

Gallons to pounds: Number of pounds (lb) = (Number of gal)(8.34 lb/gal)

Gallons to cubic feet: Number of $ft^3 = \dfrac{(\text{Number of gal})\ (1\ ft^3)}{7.48\ gal}$

Acre-feet to cubic feet: Number of ft^3 = (Number of acre-ft)(43,560 ft^3/acre-ft)

mgd to ft^3/s: Number $ft^3/s = \dfrac{(\text{Number of mgd})\ (1{,}000{,}000\ gal)\ (1\ ft^3)\quad (1\ day)\quad (1\ min)}{1\ mil\ gal\quad (7.48\ gal)\ (1{,}440\ min)\ (60\ sec)}$

ft^3/s to mgd: Number of mgd $= \dfrac{(\text{Number of } ft^3)\ (60\ sec)\ (1{,}440\ min)\ (7.48\ gal)\ \ (1\ mil\ gal)}{sec\qquad min\qquad day\qquad ft^3\quad 1{,}000{,}000\ gal}$

ft^3/s to gpd = Number of gpd = (Number of ft^3/s)(86,400 s/d)(7.48 gal/ft^3)

gpm to ft^3/s: Number of $ft^3/s = \dfrac{\text{Number of gpm}}{(60\ s/min)(7.48\ gal/ft^3)}$

ppm to Percent: Percent solution $= \dfrac{(\text{Known ppm})\quad (1\%)}{10{,}000\ ppm}$

Gallons to liters: Number of liters = (Number of gal)(3.785 L/1 gal)

Composting Equations

$$\text{Mixture's \% moisture} = \frac{[(\text{Sludge, lb})(\text{\% moisture}) + (\text{Compost, lb})(\text{\% moisture})]100\%}{\text{Sludge, lb} + \text{Compost, lb}}$$

Percent solids BC =

$$\frac{(\text{Sludge, yd}^3)(\text{lb/yd}^3)(\text{\% solids, sludge}) + (\text{Sludge, yd}^3)(\text{MR})(\text{lb/yd}^3)(\text{\% solids, chips})(100\%)}{(\text{Sludge, yd}^3)(\text{lb/yd}^3) + (\text{Sludge, yd}^3)(\text{Mix ratio})(\text{lb/yd}^3)}$$

$$\text{Compost cycle, days} = \frac{(\text{Site capacity, yd}^3)(\text{Density of compost, lb/yd}^3)}{x\ \text{Wet compost, lb/d}}$$

Cycle time, days =

$$\frac{(\text{Capacity, yd}^3)(\text{Bulk density of compost lb/yd}^3)}{\dfrac{x\ \text{Dry solids, lb/d}}{\text{Percent solids}} + \dfrac{(x\ \text{Dry solids, lb/d})(\text{MR})}{\text{Percent solids}}\dfrac{(\text{Bulk density of wood chips, lb/yd}^3)}{(\text{Bulk density of wet sludge lb/yd}^3)}}$$

Percent moisture in mixture =

$$\frac{[(\text{DB, lb/d})(\text{Percent moisture DB}) + (\text{Compost lb/d})(\text{Percent moisture compost})]100\%}{\text{DB, lb/d} + \text{Compost, lb/d}}$$

Density Equations

Density = Mass/Volume

Number of g/cm^3 = (Number of lb/gal)(454 g/1 lb)(1 gal/3,785 cm^3)

Detention Time Equation

$$\text{Detention time, hr} = \frac{\text{Volume, gal}}{\text{Flow rate, gph}}$$

$$\text{Detention time, hr} = \frac{(\text{Volume, gal})(24\ \text{hr/d})}{\text{Flow, gpd}}$$

Dewatering Formulas

Total nonfilterable residue, mg/L = Total residue, mg/L – Total filterable residue, mg/L

$$\text{Filter yield, lb/hr/ft}^2 = \frac{\text{(Wet cake flow, lb/hr)(Percent solids/100\%)}}{\text{Area, ft}^2}$$

$$\text{Vacuum filter loading, lb/d/ft}^2 = \frac{\text{(Biosolids, gpd)(8.34 lb/gal)(Percent solids)}}{\text{Vacuum filter area, ft}^2}$$

Digester Gas Production Formula

$$\text{Gas produced, ft}^3\text{/lb VS destroyed} = \frac{\text{Gas production, ft}^3\text{/d}}{\text{VS destroyed, lb/d}}$$

$$\text{Gas produced, m}^3\text{/lb VS destroyed} = \frac{\text{(Gas production, ft}^3\text{/d)}}{\text{(VS destroyed, lb/d)(35.3 m}^3\text{/ft}^3\text{)}}$$

Digester Loading Rate Equation

$$\textbf{Digester loading rate, lb VSA/d/ft}^3 = \frac{\text{lb VSA}}{\text{Volume of digester, ft}^3}$$

Where VSA = Volatile Solids Added

Digester loading, lb VS/d/ft^3 =

$$\frac{\text{(Flow, gpd)(8.34 lb/gal)(sp gr)(Percent sludge)(Percent volatile solids)}}{\text{(0.785)(Diameter)}^2\text{(Sludge level)}}$$

Digester loading, lb VS/d/1,000 ft^3 =

$$\frac{\text{(Flow, gpd)(8.34 lb/gal)(sp gr)(Percent sludge)(Percent volatile solids)}}{\text{(0.785)(Diameter)}^2\text{(Sludge level)}}$$

Digester Volatile Solids Ratio Formula

$$\text{Digester VS ratio} = \frac{\text{VS added lb/d}}{\text{lb VS in digester}}$$

$$\text{Digester VS ratio} = \frac{\text{VS added lb/d}}{\text{(lb VS in digester)(TS \%/100\%)(VS \%/100\%)}}$$

Dilution Triangle

Concentration (Conc.)$_1$ Conc.$_2$ – Conc. Desired = Number of parts Conc.

Conc. Desired

Conc.$_2$ Conc.$_1$ – Conc. Desired = $\frac{\text{Number of parts}}{\text{Total number of parts}}$

$$\frac{(\text{Number of gallons})(\text{Number of parts of Conc.}_1)}{\text{Total number of parts}} = \text{Number of gallons of Conc.}_1 \text{ required}$$

$$\frac{(\text{Number of gallons})(\text{Number of parts of Conc.}_2)}{\text{Total number of parts}} = \text{Number of gallons of Conc.}_2 \text{ required}$$

Dissolved Air Flotation: Air Rate Flow Equations

Air, lb/d = (Air flow, ft^3/min)(1,440 min/d)(0.0807 lb/ft^3, Air)

Air, lb/d = (Air flow, ft^3/min)(60 min/hr)(0.0807 lb/ft^3, Air)

Dissolved Air Flotation: Air-to-Solids Ratio Equation

$$\text{Air-to-solids ratio} = \frac{(\text{Air flow, ft}^3/\text{min})(\text{Air, lb/ft}^3)}{(\text{gpm})(\text{Percent solids}/100\%)(8.34 \text{ lb/gal})}$$

Dissolved Air Flotation: Thickener Solids Loading Equations

$$\text{Solids loading, lb/d/ft}^2 = \frac{(\text{WAS, mg/L})(\text{mgd})(\text{Sludge, lb/gal})}{\text{DAF area, ft}^2}$$

$$\text{Solids loading, lb/hr/ft}^2 = \frac{(\text{WAS, mg/L})(\text{mgd})(\text{Sludge, lb/gal})}{(\text{DAF area, ft}^2)(24 \text{ hr/d})}$$

Dosage Formulas

Chlorine dose = Chlorine demand + Chlorine residual

Chemical feed, lb/d = (Flow, mgd)(Dosage, mg/L)(8.34 lb/gal) or rearranging to solve for dosage:

$$\text{Dosage, mg/L} = \frac{\text{lb/d}}{(\text{mgd})(8.34\ \text{lb/gal})}$$

$$\text{lb/d} = \frac{(\text{mgd})(\text{Dosage, mg/L})(8.34\ \text{lb/gal})}{(\text{Percent purity}/100\%)}$$

Above formula used when the purity of a substance or solution is less than 100%.

(mgd)(x, mg/L)(8.34 lb/gal) = (mgd)(Dosage, mg/L)(8.34 lb/gal)

Where x = unknown mgd

$$\text{Chlorine, lb/d} = \frac{(\text{Dosage, mg/L})(\text{mgd})(8.34\ \text{lb/gal})}{\text{Percent available chlorine}/100\%}$$

$$\text{Chemical dosage, mg/L} = \frac{(\text{mL/min})(1{,}440\ \text{min/d})(\text{Chemical, lb/gal})}{(3{,}785\ \text{mL/gal})(\text{mgd})(8.34\ \text{lb/gal})}$$

$$\text{Chemical dosage, mg/L} = \frac{(\text{mL/min})(1{,}440\ \text{min/d})(\text{Chemical, lb/gal})}{(3{,}785\ \text{mL/gal})(\text{mgd})(8.34\ \text{lb/gal})(\text{Percent Polymer})}$$

Dry Chemical Feed Settings

$$\text{Chemical, lb/d} = \frac{(\text{Number of g/min})(1{,}440\ \text{min/d})}{454\ \text{g/lb}}$$

Extrapolation: Used for Pipes Not Flowing Full

$$\text{Division Factor} = \frac{(\text{High d/D} - \text{Low d/D})}{\text{Ratio d/D}}$$

Food/Microorganism Ratio Formula

$$\text{F/M ratio} = \frac{(\text{BOD}_5\text{, mg/L})(\text{Flow, mgd})(8.34 \text{ lb/gal})}{(\text{mg/L MLVSS})(\text{Volume in tank, mil gal})(8.34 \text{ lb/gal})}$$

or as follows since 8.34 lb/gal in the numerator and denominator cancel each other out.

$$\text{F/M ratio} = \frac{(\text{BOD}_5\text{, mg/L})(\text{Flow, mgd})}{(\text{mg/L MLVSS})(\text{Volume of tank, mil gal})}$$

Flow Rate Equations

Flow = Volume/Time

Q (Flow) = (Area)(Velocity) Example: Q, flow in ft^3/s = (Area, ft^2)(Velocity in feet per sec)

Flow in a pipe that changes size: (Area$_1$, ft^2)(Velocity 1, ft/s) = (Area$_2$, ft^2)(Velocity 2, ft/s)

$$\text{Flow } (Q)\text{, ft}^3\text{/s} = \frac{(w_1 + w_2)}{2}(\text{Depth, ft})(\text{Velocity, ft/s})$$

Gravity Thickener Solids Loading Formula

$$\text{Solids loading, lb/d/ft}^2 = \frac{(\text{Flow, gpm})(1{,}440 \text{ min/d})(\text{Percent solids})}{(\text{Gravity thickener area})(100\%)}$$

$$\text{Solids loading, lb/d/ft}^2 = \frac{(\text{Flow, gpd})(\text{Percent solids})}{(\text{Gravity thickener area})(100\%)}$$

Grit Removal Formula

$$\text{Grit removal, ft}^3\text{/mil gal} = \frac{\text{Number of gallons removed}}{(7.48 \text{ gal/ft}^3)(\text{mil gal treated})}$$

Hydraulic Digestion Time Equation

$$\text{Digestion time, days} = \frac{\text{Number of gallons}}{\text{Influent sludge flow, gal/d}}$$

$$\text{Digestion time, days} = \frac{(0.785)(\text{Diameter})^2 (\text{Depth, ft})(7.48 \text{ gal/ft}^3)}{\text{Influent sludge flow, gal/day}}$$

Hydraulic Loading Rate Equation

$$\text{Hydraulic loading rate} = \frac{\text{Total flow, gpd}}{\text{Surface area, ft}^2}$$

$$\text{Hydraulic loading rate, in./d} = \frac{\text{Flow, gpd}}{\text{(27,152 gal/acre-in.)(Area, acres)}}$$

Kilowatt Formulas

kW = (Number of hp)(0.746 kW/hp)

Kilowatt-hr/d = (hp)(0.746 kW/hp)(Operating time, hr/d)

kW = (Number of hp)(0.746 kW/hp)(Startup energy)

Lime Neutralization Formula

Lime, lb = (Volatile acids, mg/L)(mil gal)(8.34 lb/gal)

Mass Balance (Percent) Equation

$$\text{Percent mass balance} = \frac{\text{(Solids produced, lb/d + Solids removed, lb/day)(100\%)}}{\text{Solids produced, lb/d}}$$

Mean Cell Residence Time Equations

$$\text{MCRT, days} = \frac{\text{(MLSS, mg/L)(mil gal)(8.34 lb/d)}}{\text{SS wasted, lb/d + SS lb/d}}$$

MCRT, days =

$$\frac{\text{(MLSS, mg/L)(Aeration tank mil gal + Clarifier tank mil gal)(8.34 lb/gal)}}{\text{(WAS, mg/L)(Waste rate, mgd)(8.34 lb/gal) + (TSS, mg/L)(Flow, mgd)(8.34 lb/gal)}}$$

Mixture Formula

Percent mixture strength =

$$\frac{\text{Solution}_1\text{ gal (Available \%/100\%)} + \text{Solution}_2\text{ gal (Available \%/100\%)(100\%)}}{\text{gal of solution}_1 + \text{gal of solution}_2}$$

Nitrogen Loading Rate Equation

Nitrogen loading rate, lb/day = (Total Nitrogen, mg/L)(mgd)(8.34 lb/gal)

Nitrogen (Total) Equation

Total nitrogen (N) = Nitrate, mg/L + Nitrite, mg/L + TKN, mg/L

Operating Time Formulas

$$\text{Operating time} = \frac{\text{Treated water}}{\text{Flow rate}}$$

$$\text{Operating time, min/hr} = \frac{\text{(Flow, mgd)(Influent SS, mg/L – Effluent SS, mg/L)(100\%)}}{\text{(Sludge pump, gpm)(Percent solids)(24 hr/d)}}$$

Organic Loading Rate Equations

$$\text{Organic loading rate, lb BOD}_5\text{/d/acre} = \frac{(\text{BOD}_5\text{, mg/L)(Flow, mgd)(8.34 lb/gal)}}{\text{Surface area of pond, acre-ft}}$$

$$\text{Organic loading rate, lb BOD}_5\text{/d/1,000 ft}^3 = \frac{(\text{BOD}_5\text{, mg/L)(Flow, mgd)(8.34 lb/gal)}}{\text{Volume of trickling filter, ft}^3\text{/1,000 ft}^3}$$

Particulate and Soluble Biochemical Oxygen Demand Formulas

Particulate BOD_5, mg/L = (SS, mg/L)(K-value)

Soluble BOD_5 = Total BOD_5 – (K-value)(Total SS)

Total BOD_5 = (Particulate BOD_5)(K-value) + Soluble BOD_5

Perimeter Formulas

Circumference = π(Diameter) = (3.14)(Diameter)

Rectangle = 2(Length) + 2(Width)

Plant Available Nitrogen Formulas

PAN, mg/L = [MR(TKN – NH_3)] + [0.50(NH_3)] + (NO_3 + NO_2)

$$\text{PAN, dry tons/acre} = \frac{\text{Plant nitrogen required, lb/acre}}{\text{PAN, lb/dry ton}}$$

PAN, lb/dry ton = [(Organic N, mg/kg)(MR) + (Ammonia N, mg/kg)(VR)](0.002 lb/dry ton)

Population Equivalent Equation

$$\text{Number of people} = \frac{(\text{BOD}_5\text{, mg/L})(\text{mgd})(8.34\ \text{lb/gal})}{\text{lb/d of BOD}_5\text{/person}}$$

Population Loading Equation

$$\text{Population loading, people/acre} = \frac{\text{Number of people served}}{\text{Area of pond(s), acres}}$$

Pressure Formulas

$$\text{psi} = \frac{\text{Depth, ft}}{2.31\ \text{ft/psi}}$$

Height, ft = (psi)(2.31 ft/psi)

psi = (Depth, ft)(0.433 psi/ft)

$$\text{Pressure} = \frac{\text{Force, lb}}{\text{Area, ft}^2}$$

Pressure, lb/ft^2 = (Height or Depth, ft)(Density, 62.4 lb/ft^3)

$$\frac{\text{Pressure}_A}{w} + \frac{\text{Velocity}_A{}^2}{2g} = \frac{\text{Pressure}_B}{w} + \frac{\text{Velocity}_B{}^2}{2g}$$

Percent Mixture Formulas

$$\text{Percent mixture} = \frac{(\text{Soln.}_1, \text{lb})(\text{Avail }\%/100\%) + (\text{Soln.}_2, \text{lb})(\text{Avail }\%/100\%)(100\%)}{\text{strength Soln.}_1, \text{lb} + \text{Soln.}_2, \text{lb}}$$

Percent mixture strength =

$$\frac{[(\text{Soln.}_1, \text{gal})(\text{lb/gal})(\text{Avail }\%/100\%) + (\text{Soln.}_2, \text{gal})(\text{lb/gal})(\text{Avail }\%/100\%)]100\%}{(\text{Soln.}_1, \text{gal})(\text{lb/gal}) + (\text{Soln.}_2, \text{gal})(\text{lb/gal})}$$

Percent Recovery Equation

$$\text{Percent recovery} = \frac{\text{Cake TS, \%(Feed sludge TSS, \% – Return flow TSS, \%)(100\%)}}{\text{Feed sludge TSS, \%(Cake TS, \% – Return flow TSS, \%)}}$$

Percent Reduction Equations

$$\text{Percent VS reduction} = \frac{(\text{Influent} - \text{Effluent})(100\%)}{\text{Effluent} - (\text{Effluent})(\text{Influent})}$$

$$\text{Percent VM reduction} = \frac{(\text{Percent influent VM} - \text{Percent effluent VM})(100\%)}{[\text{Percent influent VM} - (\text{Percent influent VM})(\text{Percent effluent VM})]}$$

Percent moisture reduction =

$$\frac{(\text{Percent influent moisture} - \text{Percent moisture, after digestion})(100\%)}{[\text{Percent influent moisture} - (\text{Percent influent moisture})(\text{Percent moisture, after digestion})]}$$

Percent Removal Formula

$$\text{Percent ntu removal} = \frac{(\text{Influent ntu} - \text{Effluent ntu})(100\%)}{\text{Influent ntu}} \text{ or } \frac{(\text{In} - \text{Out})(100\%)}{\text{In}}$$

$$\text{Percent BOD}_5 \text{ removal} = \frac{(\text{Influent BOD}_5 - \text{Effluent BOD}_5)(100\%)}{\text{Influent BOD}_5} \text{ or } \frac{(\text{In} - \text{Out})(100\%)}{\text{In}}$$

$$\text{Percent removal efficiency} = \frac{(\text{Solids removed, mg/L})(100\%)}{\text{Influent solids, mg/L}}$$

Percent Strength of Solutions and Solids

$$\text{Percent strength} = \frac{(\text{Number of lb of chemical})(100\%)}{\text{Number of lb, Water} + \text{lb chemical}}$$

$$\text{Percent strength} = \frac{(\text{Number of grams of chemical})(100\%)}{\text{Number of grams, water} + \text{grams chemical}}$$

$$\text{Percent total solids} = \frac{(\text{Dry sample in grams})(100\%)}{\text{Sludge sample in grams}}$$

Percent Settled Sludge and Solids Formulas

$$\text{Percent settled sludge} = \frac{(\text{Settled sludge, mL})(100\%)}{\text{Total sample vol., mL}}$$

or similarly:

$$\text{Percent settleable solids} = \frac{(\text{Settled sludge, mL})(100\%)}{\text{Sample size, mL}}$$

$$\text{Percent inorganic solids} = \frac{(\text{Dry sample in grams})(100\%)}{\text{Sludge sample in grams}}$$

$$(2^\circ\ \text{gpd})(2^\circ\ \text{sludge lb/gal})(x\%\ 2^\circ\ \text{sludge}) = (1^\circ\ \text{sludge, gpd})(1^\circ\ \text{sludge lb/gal})(\%\ 1^\circ\ \text{sludge})$$

Where x = unknown %

Phosphate Loading Rate Equation

Phosphorus (P) loading rate, lb/d = (P, mg/L)(mgd)(8.34 lb/gal)

Pit Volume and Days-to-Fill Formulas

$$\text{Screenings, ft}^3\text{/mil gal} = \frac{\text{Number of ft}^3\text{/d}}{\text{Number of mgd}}$$

$$\text{Number of days to fill} = \frac{\text{Pit volume, ft}^3}{\text{Screenings removed, ft}^3\text{/d}}$$

Pumping and Pumping Cost Formulas

$$\text{mhp} = \frac{(\text{whp})}{(\text{Motor efficiency})(\text{Pump efficiency})}$$

Where mhp = motor horsepower and whp = water horsepower

$$\text{mhp} = \frac{(\text{Flow, gpm})(\text{TH, ft})}{(3{,}960)(\text{Motor efficiency})(\text{Pump efficiency})}$$ Where TH = Total head

$$\text{bhp} = \frac{(\text{Flow, gpm})(\text{Differential pressure, psi})}{(1{,}714)(\text{Pump efficiency})}$$

Brake hp = (hp)(Motor efficiency) Where hp = horsepower

Water hp = (mhp)(Motor efficiency)(Pump efficiency)

Water hp = (bhp)(Pump efficiency)

Motor hp = bhp/Motor efficiency

Brake hp = whp/Pump efficiency

Cost, \$/d = (Motor hp)(24 hr/d)(0.746 kW/hp)(Cost/kW-hr)

Pumping Rate Equations

Pumping rate = Flow, gal/Time, min

$$\text{Pump's discharge rate, gpm} = \frac{\text{Discharge, gal}}{\text{Time, min}}$$

Discharge rate, gpm = Influent flow, gpm + Level drop, gpm

Number of gal per stroke = (0.785)(Bore diameter, ft)^2(Stroke, ft)(7.48 gal/ft^3)

Ratios

$$\text{Ratio} = \frac{\text{Recirculated flow}}{\text{Plant influent flow}}$$

$$\frac{\text{Speed setting}_1\text{, Percent}}{\text{Polymer dosage}_1\text{, mL}} = \frac{\text{Speed setting}_2\text{, Percent}}{\text{Polymer dosage}_2\text{, mL}}$$

$$\frac{\text{Chlorine dosage}_1\text{, mg/L}}{\text{Flow}_1\text{, mgd}} = \frac{\text{Chlorine dosage}_2\text{, mg/L}}{\text{Flow}_2\text{, mgd}}$$

$$\frac{\text{Digester solids}_1\text{, lb/d}}{\text{Flow}_1\text{, gpm}} = \frac{\text{Digester solids}_2\text{, lb/d}}{\text{Flow}_2\text{, gpm}}$$

$$\frac{\text{Alum dosage}_1\text{, mL}}{\text{Speed setting}_1\text{, \%}} = \frac{\text{Alum dosage}_2\text{, mL}}{\text{Speed setting}_2\text{, \%}}$$

Screenings Formula

$$\text{Screenings, ft}^3\text{/mil gal} = \frac{\text{Number of ft}^3\text{/d}}{\text{Number of mgd}}$$

Seed Sludge Equation

$$\text{Seed sludge, gal} = \frac{\text{(Capacity of digester)(Percent seed sludge required)}}{100\%}$$

Sludge Age (Gould) Equation

$$\text{Sludge age, days} = \frac{\text{Solids under aeration, lb}}{\text{Solids added, lb/d}}$$

$$\text{Sludge age, days} = \frac{\text{(MLSS, mg/L)(Volume of aeration tank)(8.34 lb/gal)}}{\text{(SS, mg/L)(Flow, mgd)(8.34 lb/gal)}}$$

Sludge Pumping Formula

Sludge, lb/d = (Pumping, min/day)(24 hr/d)(Pump rate, gpm)(8.34 lb/gal)(sp gr of sludge)

(Primary sludge, gal)(Primary sludge, lb/gal)(Percent PSS) =

(x Thickened sludge, gal)(Thickened sludge, lb/gal)(Percent TSS)

Sludge Removed Equation

SS removed, lb/d = (SS removed, mg/L)(Number of mgd)(8.34 lb/gal)

Sludge Volume Index and Sludge Volume Density Equations

$$\text{SVI} = \frac{(\text{SS, mL})}{\text{MLSS, g/L}}$$

$$\text{SDI} = \frac{(\text{MLSS, g})(100\%)}{\text{SS, mL}}$$

Sodium Absorption Ratio Formula

$$\text{Sodium absorption ratio} = \frac{\text{Na}^{+}}{[(0.5)(\text{Ca}^{2+} + \text{Mg}^{2+})]^{1/2}}$$

Solids Balance (Digester) Equations

Total solids, lb/d = (Raw sludge, lb/day)(Percent solids)

Fixed solids, lb/d = Total solids, lb/day – VS, lb/d

Water in sludge, lb/d = Sludge, lb/day – Total solids, lb/d

$$\text{Percent VSR} = \frac{(\text{In} - \text{Out})(100\%)}{\text{In} - (\text{In})(\text{Out})}$$

Gas produced, lb/d = (Effluent VS, lb/day)(Percent VSR)

VS in digested sludge, lb/d = Influent VS, lb/d – Destroyed VS, lb/d

$$\text{Total digested solids, lb/d} = \frac{\text{VS digested, lb/d}}{\text{Percent digested VS}}$$

Fixed solids, lb/d = Total digested solids, lb/d – VS digested, lb/d

$$\text{Digested sludge, lb/d} = \frac{\text{Total digested solids, lb/d}}{\text{Digested sludge percent solids}}$$

Water in digested sludge, lb/d = Sludge, lb/d – Total solids, lb/d

Solids Loading Rate Equations

$$\text{Solids loading rate} = \frac{(\text{MLSS, mg/L})(\text{mgd})(8.34\ \text{lb/gal})}{\text{Area, ft}^2}$$

$$\text{Solids loading rate, lb/d/ft}^2 = \frac{(\text{Percent solids})(\text{Biosolids added, gpd})(8.34\ \text{lb/gal})}{(\text{Surface area, ft}^2)}$$

Solids Produced Formula

Solids produced, lb/d = (BOD_5 removed, lb/d)(0.85 lb solids/lb BOD_5)

Solids Pumping Equations

$$\text{Solids, lb/d} = \frac{(\text{Sludge, lb/d})(\text{Percent solids})}{100\%}$$

Solids, lb/d = (Pumping, min/d)(24 hr/d)(Pump rate, gpm)(8.34 lb/gal)(sp gr of sludge)

Solids, lb/d = (Time, min/cycle)(cycles/d)(Pump rate, gpm)(8.34 lb/gal)(Percent solids)

Solids Retention Time Equation

Target SRT =

$$\frac{(\text{MLSS mg/L})(\text{Clarifier, Aerator Volume, mil gal})(8.34\ \text{lb/gal})}{(\text{RAS SS mg/L})(x\ \text{mgd})(8.34\ \text{lb/gal}) + (\text{Effluent SS, mg/L})(\text{Flow, mgd})(8.34\ \text{lb/gal})}$$

Where x = unknown mgd

Solids Under Aeration

Number of lb, solids = (SS, mg/L)(Number of mil gal)(8.34 lb/gal)

Solution Formulas

$(\text{Concentration}_1)(\text{Volume}_1) + (\text{Concentration}_2)(\text{Volume}_2) = (\text{Concentration}_3)(\text{Volume}_3)$
or
abbreviating the above equation: $C_1V_1 + C_2V_2 = C_3V_3$

$$\text{Percent HTH solution} = \frac{(\text{lb HTH})(100\%)}{(\text{Number of gal})(8.34\ \text{lb/gal})}$$

$(\text{Solution}_1, \text{percent})(x\ \text{gal}, \text{Solution}_1)(\text{Solution}_1, \text{lb/gal}) =$

$(\text{Solution}_2, \text{percent})(\text{Solution}_2, \text{gal})(\text{Solution}_2, \text{lb/gal})$

Specific Gravity Equations

Specific gravity (sp gr) = Density of substance/Density of water

$$\text{sp gr} = \frac{\text{Solute, lb/gal}}{8.34\ \text{lb/gal}}$$

$$\text{sp gr} = \frac{\text{Number of lb/ft}^3}{62.4\ \text{lb/ft}^3}$$

Statistic Formulas

$$\text{Average} = \frac{\text{Sum of all measurements}}{\text{Number of measurements}}$$

or

$$\text{Arithmetic mean} = \frac{\text{Sum of all measurements}}{\text{Number of measurements}}$$

Median is the middle value.

Range = Largest value – Smallest value

Mode is the value that occurs most frequently. Note: There also can be two or more modes within a set of numbers.

Geometric mean = $[(X_1)(X_2)(X_3)(X_4).....(X_n)]^{1/n}$

Standard deviation = $[\Sigma f(X - \overline{X})^2/n - 1]^{1/2}$

Surface Loading Formula

$$\text{Surface loading rate} = \frac{\text{gallons per day (gpd)}}{\text{Number of ft}^2}$$

Suspended Solids Loading Equation

Suspended solids, lb/d = (SS, mg/L)(Number of mgd)(8.34 lb/gal)

Temperature Formulas

°C = 5°C /9°F (°F–32)

°F = 9°F /5°C (°C) + 32°F

Total Force and Hydraulic Press Equations

Total force, pounds = (Area, in.^{2})(psig) or Total force = (Pressure)(Area)

$$\text{Pressure} = \frac{\text{Total force, lb}}{\text{Area, ft}^2}$$

Total Head Formulas

Total head, ft = Total static head, ft + Head losses, ft

$$\text{TDH} = \frac{\text{(Differential pressure)(2.31 ft/psi)}}{\text{Specific gravity}}$$

Vacuum Filter Operating Time Equation

$$\text{Filter yield, lb/hr/ft}^2 = \frac{\dfrac{\text{(Solids, lb/d)(Percent recovery)}}{\text{(Filter operation, lb/d)(100\%)}}}{\text{Filter area ft}^2}$$

Velocity Equations

$$\text{Velocity, ft/s} = \frac{\text{Flow, gpm}}{(\text{Width, ft})(\text{Depth, ft})(60 \text{ sec/min})(7.48 \text{ gal/ft}^3)}$$

or

$$\text{Velocity, ft/s} = \frac{\text{Flow, gpm}}{(\text{Area, ft}^2)(60 \text{ sec/min})(7.48 \text{ gal/ft}^3)}$$

Volatile-Acids-to-Alkalinity Ratio Formulas

Ratio = Volatile acids/Alkalinity

Volatile acids = (Alkalinity)(Ratio)

$$\text{Digester VS ratio} = \frac{\text{VS added lb/d}}{\text{lb VS in digester}}$$

Volatile Solids Destroyed Equation

$$\text{VS destroyed, lb/d/ft}^3 = \frac{(\text{Flow, gpd})(\text{Sludge, lb/gal})(\text{SSC, \%})(\text{VSC, \%})(\text{VSR, \%})}{\text{Digester capacity, 1,000 ft}^3}$$

Volatile Solids: lb/d (Percent)

VS, lb/d = (Number of lb/d, sent to digester)(Percent VS/100%)

$$\text{VS, lb/d} = \frac{(\text{Number of gpd to digester})(\text{Percent solids})(\text{Percent VS})(8.34 \text{ lb/gal})}{100\% \qquad 100\%}$$

Volatile Solids Pumping Formula

VS, lb/d =

(Time, min/cycle)(cycles/day)(Pump rate, gpm)(8.34 lb/gal)(Percent, solids)(Percent VM)

Volume Equations

Volume of a basin in ft^3 or m^3 = (Length)(Width)(Depth)

Volume of a basin in gallons = (Length)(Width)(Depth)(7.48 gal/ft^3)

Volume of a cone in ft^3 or m^3 = $1/3\pi r^2$(Height or Depth)

Volume of a circular tank in ft^3 or m^3 = πr^2(Height) or Volume of a pipe = πr^2(Length) or

Volume of a cylindrical tank in ft^3 or m^3 = $(0.785)(\text{Diameter})^2(\text{Height})$

$$\text{Volume of a trough in gallons} = \frac{(b_1 + b_2)}{2}(\text{Depth of water})(\text{Length})(7.48\ \text{gal/ft}^3)$$

Where b_1 = bottom width of trough or ditch and

b_2 = surface width of trough or ditch

$$\text{Volume, gal} = \frac{(\text{Length}_1 + \text{Length}_2)}{2}\frac{(\text{Width}_1 + \text{Width}_2)}{2}(\text{Depth, ft})(7.48\ \text{gal/ft}^3)$$

$$\text{Volume of sphere, ft}^3 = \frac{4\pi r^3}{3}$$

Digester capacity, ft^3 = $\pi(\text{radius})^2$(Height, ft)

Equation below is for a partially filled pipe division factor for determining flow:

$$\text{Division factor} = \frac{(\text{High d/D} - \text{Low d/D})}{\text{Ratio d/D}}$$

Waste Activated Sludge Loading Rate Equation

WAS, lb/d = (WAS, mg/L)(Number of mgd)(8.34 lb/gal)

Waste Activated Sludge Pumping Rate Formula

$$\text{Number of mgd} = \frac{\text{WAS, lb/d}}{(\text{Number of mg/L WAS})(8.34\ \text{lb/gal})}$$

Waste Rate Equation

$$\text{Waste rate, mgd} = \frac{\text{Waste, lb}}{(\text{WVS concentration, mg/L})(8.34\ \text{lb/gal})}$$

The following two equations are used to find the waste rate (above equation).

$$\text{Desired MLVSS, lb} = \frac{(\text{Primary effluent COD, mg/L})(\text{mgd})(8.34\ \text{lb/gal})}{\text{Desired COD lb/MLVSS lb}}$$

$$\text{Existing MLVSS, lb} = (\text{MLVSS, mg/L})(\text{Aeration tank, mil gal})(8.34\ \text{lb/gal})$$

$$\text{Waste rate, gpd} = \frac{(\text{Solids produced, lb/d})(1{,}000{,}000/\text{M})}{(\text{Waste TSS, mg/L})(8.34\ \text{lb/gal})}$$

Waste rate, lb/d =

$$\frac{\text{MLSS, mg/L[AT, mil gal} + \text{CT, mil gal]}(8.34\ \text{lb/gal})}{\text{Desired MCRT}} - (\text{TSS, mg/L})(\text{Flow, mgd})(8.34\ \text{lb/gal})$$

Weir Length Formula

$$\text{Weir length, ft} = \pi(\text{Diameter, ft})$$

$$\text{Weir length, ft} = 2\pi(\text{radius, ft})$$

Weir and Surface Overflow Rate Equations

$$\text{Weir overflow rate} = \frac{\text{Flow, gpd}}{\text{Weir Length, ft}}$$

$$\text{Surface overflow rate} = \frac{\text{Flow, gpd}}{\text{Area, ft}^2}$$

Length

1 inch (in.) = 2.54 centimeters (cm)

100 cm = 1 meter (m)

1 m = 39.37 in.

1 m = 3.281 feet (ft)

1 yard = 0.9144 m

1,000 m = 1 kilometer (km)

1 km = 1.609 miles (mi)

This page intentionally blank.

www.ingramcontent.com/pod-product-compliance
Lightning Source LLC
Chambersburg PA
CBHW081125180926
51611CB00004B/58
9781647170097